Art as Inquiry

American University Studies

Series XX
Fine Arts

Vol. 32

PETER LANG
New York • Washington, D.C./Baltimore
Bern • Frankfurt am Main • Berlin • Vienna • Paris

Marga Bijvoet

Art as Inquiry

Toward New Collaborations Between Art, Science, and Technology

PETER LANG
New York • Washington, D.C./Baltimore
Bern • Frankfurt am Main • Berlin • Vienna • Paris

Library of Congress Cataloging-in-Publication Data

Bijvoet, Marga.
Art as inquiry: toward new collaborations between art, science, and technology/ Marga Bijvoet.
p. cm. — (American university studies. XX, Fine arts; v. 32)
Includes bibliographical references and index.
1. Art and technology. 2. Art and science. 3. Art, Modern—20th century.
I. Title. II. Series: American university studies. Series XX, Fine arts; vol. 32.
N72.T4B54 701'.05'09045—dc20 96-33375
ISBN 0-8204-3382-9
ISSN 0890-421X

Die Deutsche Bibliothek-CIP-Einheitsaufnahme

Bijvoet, Marga:
Art as inquiry: toward new collaborations between art, science, and technology/ Marga Bijvoet. –New York; Washington, D.C./Baltimore; Bern; Frankfurt am Main; Berlin; Vienna; Paris: Lang.
(American university studies: Ser. 20, Fine arts; Vol. 32)
ISBN 0-8204-3382-9
NE: American university studies/ 20

The paper in this book meets the guidelines for permanence and durability of the Committee on Production Guidelines for Book Longevity of the Council of Library Resources.

Printed in the United States of America.

Contents

PART II

Acknowledgments

Late 1979 I arrived in New York on a Fulbright-Hays exchange scholarship from the Netherlands America Council for Educational Exchange to begin research on Environmental Art and how it had been dealt with in American art criticism. Having met a number of artists working in the field a few years before during the organization of an exhibition whereby artists were invited to do a project for a specific site, I began my research by diving into New York's art world. I am still grateful for the help offered by many people who introduced me to the institutions necessary for my research as well as the 'New York art scene.' My interest in media art soon led to an active involvement in this community, as a curator and writer. After I had moved to Los Angeles my research gradually shifted - surely influenced by the 'climate' there - toward the so-called 'new physics,' Eastern philosophy, and possible relationships with the visual arts. I remember many discussions with artists on this topic. Yet, how could one possibly measure such an influence? Continued research into environmental art forms and media arts finally gave me the idea that the starting points of artists who turned to nature and those who utilized technology in the visual arts might not be as diametrically opposed as supposed, both having as their common bond the ideals of the sixties "to break the boundaries of art" and the search for new forms and structures for art.

There is a general belief that not much came of the ideals of the sixties. After all, the majority of the artists who tried to flee the art world were in the end co-opted and returned to making objects. Yet, some artists who started their careers at that time kept investigating the natural environment and new technologies, inspiring a 'second generation' of artists working in public places and digital, electronic media; predominantly outside the mainstream of the art world. What happened? So, how these artists continued to develop their idea(l)s since the late sixties into a new framework by transforming the form of the art object and the function of the artist ultimately became the focal point of my concerns. The result is not a survey, nor presents a linear and chronological development. It intends to delineate aspects and ideas that seem crucial to the investigations of this period, and the development into the current 'movements' of Art in Public Places and Media Art.

The present edition is an extended and up-dated version of a dissertation which was presented for a doctorate degree at the Erasmus University of Rotterdam, Netherlands in 1994. I am very grateful to Dr. W.A.L. Beeren and Dr. H.W.van Os, who inspired me tremendously when I was a student at the University of Groningen, Netherlands and who have unknowingly set the directions of my future studies in contemporary art. This book would not have been possible without the support of the many artists who have gratuitously given me their time for interviews or just to exchange thoughts and ideas.

Many thanks go to artists Jacki Apple, Robert and Lynn Bassler, Michael Brewster, Carl Cheng, Lloyd Hamrol, Helen and Newton Harrison, Anna Homler, Nancy Holt, Robert Irwin, Maria Nordman, Sheila Pinkel, Elisabeth Richardson, Rachel Rosenthal, Paul Ryan, Barbara Smith, Alan Sonfist, the Vasulkas, Bill Viola, and Ruth Weisberg, and many other friends not mentioned here. Special thanks also go to Joan Hugo, and Kathy Rae Huffman, who have supported and stimulated me during all these years. Last but not least, I am very grateful for the critical reading of the text of my dissertation by professor H.W. Crouwel and Antje von Graevenitz, and above all for the final and scrutinous reading of the manuscript by long-time friend Rudy Bremer and Wolfgang Preikschat who helped design the final lay-out.

PART I

Introduction

Moving Out

In general the second half of the sixties can be characterized as an explosion of creativity, in which artists explored new means of expression, new materials, new forms. The key word at the time was that artists wanted to "break the boundaries," or "cross the boundaries" of the well-defined art object. One strategy to do so was to bring cheap, non-precious daily-life and junk objects, natural materials like dirt or plants, and temporary materials subject to decay into the exhibition space. Names like Fluxus, Arte Povera, Process Art were introduced to name these experiments. The next step was to go outdoors, into nature or the city environment. Among the first artists to leave the studio were the so called "earth artists." Michael Heizer went into the desert. So did Walter De Maria. Robert Smithson had a preference for deserted industrial wastelands. In the beginning the Earth Art movement was generally perceived as an anti-art establishment statement, whereby artists were now using the land as their canvas or as sculptural material. Heizer has voiced his opinion about its origins: "One of the implications of earth art might be to remove completely the commodity status of a work of art and allow a return to the idea of art as more of a religion."[1] So, in addition artists objected to the conventional triangular artist-gallery-museum system, because it had become too commercial and corrupt in the eyes of many artists. The elevated tone of Smithson's voice against the art system leaves no room for misunderstanding: "Visiting a museum is a matter of going from void to void. Hallways lead the viewer to the things once called 'pictures' and 'statutes.' Anachronisms hang and protrude from every angle. Themes without meaning press on the eye. Multifarious nothings permute into false windows (frames) that open up onto a veracity of blanks," he wrote in his article "Some Void Thoughts on Museums."[2]

Yet the initial negative attitude turned positive with Robert Smithson's site/non-site dialectic which became the theoretical foundation for a series of environmental works, relating internal and external aspects of the chosen site, such as the geology and human history (as exemplified by human destruction, for example), which eventually led to his thoughts on land reclamation as a future possible function of art. This quotation already shows that the flight outdoors did not just remain an attack on the art system. The curator of the first *Earth Art* exhibition (1967) at the Andrew Dickson White Gallery, Cornell University, Ithaca NY, may have recognized this early on, when he wrote about art's being eventually reintegrated into the social system rather than remaining something distinct and remote from other activities. Once the transition to a socially integrated art is complete, we may see the full implementation of an art impulse in an advanced technological society. Earth artists just may fulfill an ideal stated earlier by John Cage to "set forth a view of the arts which does not separate them from the rest of life, but rather confuses the difference between Art and Life, just as it diminishes the distinctions between space and time."[3]

Into Technology

The wave of new technologies that swept the sixties could not but affect the visual arts. It was the time that the electronic media and the computer gradually became available to the public. Among the new materials explored by artists initially were video, laser, holography, the computer. As it happens, these media are characterized by immateriality, temporalness, and non-objectiveness, and although expensive production modes, they are in essence non-precious. In fact, one of the major technical qualities of video or computer works is their reproducibility. The field of presentation of these media - being by nature information and communication media - was initially often seen to lie outside the museum and gallery walls: on television, festivals, or in the context of plain community services and education.

Thus, around 1965/1966 there was a situation in which a group of New York artists expressed a growing interest in technology, looking for access and knowledge. When re-reading the writings in publications, newspapers, magazines one notices an incredibly optimistic belief in this technological progress. During the twentieth century there had already been a number of art movements that were characterized by the attention that was paid to the artistic possibilities of the latest techniques, following current scientific discoveries (Bauhaus, Constructivism). In the latter half of the sixties there was yet another peak, which happened to be congruent with the introduction of the technologies that have now permeated our daily existence. The heyday of this Art and Technology Movement lasted only four years (c.1968 until 1972). The relationship between art and technology and art and science was a much debated topic, but was mostly discussed in terms of similarities (creativity) and differences (methods and objectives). In 1966, writer John Gruen wrote enthusiastically of "artists and engineers deliberately joining forces. ... Their aim is to start a revolution. To overthrow old concepts, to reach into the unknown and produce art works that will combine the most advanced technological discoveries with the most daring, the most outrageous creative ideas an artist may be capable of dreaming up."[4] Two years later, artist/writer Douglas Davis began an article euphorically with: "Living as they do, in a super technological society, American artists have quite naturally turned to the products, processes and imagery of science and industry. Some approach technology with traditional attitudes, others are using it to alter the very definition of art, but all who succumb to its fascination have responded with a new sense of exhilaration and discovery."[5] It could have been just another of those art and science/technology waves that rippled the arts in the twentieth century, but even at the time the Art and Technology Movement was already perceived as being different from the previous ones.

In Search of Another Context

Sometimes it seems that there have been as many expressions in the visual arts as artists since the sixties. Yet the 'moving out' into nature or the environment and 'into technology' since the mid-sixties are the two movements which stand out. The first went into history as the Earth Art and Environmental Art movements, the second as the Art and Technology movement, of which video and computer art became the best known exponents.

The Earth Art and Environmental Art, and the Art and Technology movements are said to have started from the premise "to break the boundaries of art," to change the commercial art world structure. In review this is only partially true. Even if both 'movements' only existed a few years, they have initiated new thoughts about the function of art and the role of the artist. The changes that took place in the work of a number of artists since the mid-sixties in the United States and elsewhere might be interpreted as a beginning of a re-orientation; as the germination of a search for another context, or function. The initial purpose to break the boundaries of art and the art system did not stop at enlarging the boundaries by including other territory. What happened was that the artists who went into the public environment sought contacts with and access to other disciplines to create a work of art that would be a part of the environment, so that the work might function through a relationship with or in context with 'the real world.' For the first time in a long time artists who ventured to investigate new media began to seek an equal collaboration with engineers and technicians, even if it was initially the only road which led to the knowledge and access of the new information and communication technologies, and born out of necessity. The positive aspect in the exploration of new structures was directed to ideally include art and artist in a new social system. This ideal in fact connected the new artistic ventures with the utopian and often rather vague notions of social and political change envisioned in the 1960s.

Indeed, the most important characteristic of the technologically oriented artists was a new type of interdisciplinary collaboration between artists, scientists and technicians. Later on inter- and cross-disciplinary collaborations also became normal in the production of public environmental sculpture. This implied that the artist entered into a new relationship with the environment, space, public arena, onto the terrain of other sciences. The important thing is that the art work became a part of a larger context, that this contextualization of the art work became the starting point for a number of artists to create a work that was no longer an object, but one that consisted of elements that were related to one another as in a system.

Although there is no specific correlation between the developments in the visual arts and the sciences, it is noteworthy that the shifting interests in the arts also revolved around concepts of time and space as a time-bound reality (Albert Einstein), events and processes, used with implied randomness and a probability factor (Werner Heisenberg). At that time the ideas of a systems approach and cybernetics spread rapidly and found application in numerous disciplines, both in the natural and social sciences. A central element in both theories was the development of a (mathematical) language which would facilitate an understanding between different scientific disciplines, and interdisciplinarity. The descriptive language used here found its way into the brochures describing the features of new technologies like video and the computer (feedback, closed-circuit, random access memory, etc.). In addition, "seeing things in relations" - a concept borrowed from systems analysis - became an important phrase. I think it is important to recall that the new technologies were basically technical applications of a number of preceding scientific discoveries with such penetrating consequences as to force scientists to completely rethink the mechanical world model on which their research was hitherto based.

At any rate, the discoveries of two sciences were central in this constellation: general system theory and cybernetics. Through the writings of Norbert Wiener, Herbert Marshall McLuhan and R. Buckminster Fuller they also infiltrated into the art world. In particular those artists informed themselves who showed an inclination and curiosity in new technologies, like video, or computer graphics. Marshall McLuhan's *Understanding Media* was widely read, and his "the medium is the message" became an adage, even if it was not always precisely understood.

Although it might seem at first sight that there is no continuity between the Environmental Art and the Art and Technology movements of the late sixties and the nineties' developments in Art in Public Places and the Media Arts, which now include the new Virtual Reality and Cyber Arts, both have their roots in this period, for it is the search for a new context which connects the two periods. And although at first sight oppositional movements which might even work against each other, it is this search for a new *context* in which the arts could function differently and which would in turn involve a changing role of the artist, which constitutes a rarely discussed but central element in the development in both directions. How artists ventured onto new terrains and what this meant in terms of changing the conditions for the production, presentation and reception of the art works will be the central theme of this book. Further I will discuss the far-reaching consequences this had for the traditional analysis of the art work, which was predominantly based on style and form.

Contemporary Art Criticism

Theoretically, the avant-garde of the twentieth century still adhered to the concept of the autonomous art object, which had been the dominant mode since the Romantic period. Initially this brought with it possibilities for the artist to free himself of the religious or secular (iconographic or other) obligations laid upon him by his patrons. Together with this development evolved a social position wherein the artist was not only independent, but also obtained a status different from other professional occupations. In the twentieth century, the word 'autonomous' became defined by the artist's possibility to decide freely about any form or material. The avant-garde was born and with it the obsession to have to continuously change and create something new to stay ahead, as the rapid sequence of movements in this century proves. The linear historical interpretation perceives this succession of styles in terms of action and reaction. Thus the tendency of the 1960s to renounce the art object in favor of non-objective materials, such as light, sound, video, natural processes, and so on was considered a 'natural' reaction to, for example, the minimalist works, based on geometric precision, or to the large object sculpture.

Yet, the new developments in art confronted both the spectator and critic/curator with an infinite variety of configurations that could hardly be described only stylistically or formally. Indeed, the multiplicity of directions in the visual arts have conveniently been defined as pluralist or postmodern. These terms in themselves do not mean very much except indicating 1. an overall tendency toward undefined diversity and 2. a 'break' with the previous period called modernism. Other terms to circumscribe the qualities of these media were introduced, such as intermedia, anti-form, dematerialization. When Jonas Mekas, filmmaker and critic, wrote enthusiastically "suddenly the intermedia shows are all over town," referring to the light shows, multiple film and slide projections, and light and motion art forms that held the attendence of a group of artists, in 1966, he seems to have felt intuitively that the traditional language could no longer be used.[6] The so-called anti-form types of art included works in which the materials were subjected to processes, or actions like piling, hanging, stacking; works that investigated the properties of the materials themselves. However, the often used term antiform was at best an indication of an attitude of denunciation. The materials were characterized by a non-precious, daily-life and often temporary quality, the research was characterized by elements of chance, randomness and indeterminacy.

Lucy R. Lippard and John Chandler introduced the term "dematerialization," coming from physics, to circumscribe such ephemeral concepts as time and space, ideas as art, invisible systems, and so forth.[7] (The term dematerialization had been introduced into the sciences to denote the continuing mathematization of research into invisible structures of nature and reality.) One may argue that many of these aspects (such as time, process, randomness and chance, the relationships to environment and space) have correctly been mentioned by art critics and art scholars to explain in part the new directions of the times. However, the methods of interpretation generally remained within a concept of art history as a linear development of styles evolving from each other as reaction and opposition, or as a continuous change of realistic and abstract modes. Among the few exceptions who perceived the need for another theoretical approach was the English writer/critic Lawrence Alloway. Already in 1966, Alloway wrote an article titled "Systemic Painting," in which he set out to outline a general theory denoting the use of systems in abstract painting, and in particular 'field' painting. While critically surveying the art-historical and critical approaches of such distinguished contemporary art writers as Harold Rosenberg, Irving Sandler, Barbara Rose, Meyer Schapiro, Clement Greenberg, among others, he himself proposed the introduction of a number of notions from system theory, such as process, pattern, field, time-space organization, dynamics, and to consider the relationship of the part to the whole.

Alloway wanted his theory of interpretation to reach beyond the formal and stylistic boundary of the art object, which he thought to have become a constricted "web of formal relations," as opposed to "a system (which) is an organized whole, the parts of which demonstrate some regularities. A system is not antithetical to the values suggested by such art world word-clusters as humanist, organic and process. On the contrary, while the artist is engaged with it, a system is a process."[8] Willoughby Sharp, who was the editor of the avant-garde magazine *Avalanche* (published from 1970-1976) and one of the main protagonists of the intermedia art movements in the late sixties, was also one of the first writers who discussed art works that used new materials and methods in a terminology that derived from the systems sciences.[9] Although Sharp's essays come across as an unsystematic accumulation of systems terminology, he saw a connection between the ideas upon which the new science of systems analysis (initially developed by Ludwig von Bertalanffy, see below) was based and the new Earth Art. His essay for the *Earth Art* exhibition at Cornell University, "Notes Towards an Understanding of Earth Art," was an attempt to describe these works as functions of a system. While Sharp did not further develop his ideas theoretically in this direction, Alloway introduced the notion of art as message, whereby he used the definition from cybernetics.[10] He held the point of view that "Cross-references in the arts now need an idea of sufficient generality to include, but not to discard the contribution of visualist art criticism. The theory of messages as proposed by Norbert Wiener (*The Human Use of Human Beings,* 1950) satisfies this condition: messages are themselves a form of pattern and organization. Thus the formal structure and the significative function of a work of art are comparable as pattern. Instead of opposing purely pictorial form to a 'literary' cargo, both aspects of art are comfortable to one idea of organization."[11] In retrospect, Lawrence Alloway has turned out to be the only writer who continued to develop this direction of interpretation for a while, e.g. in his writings on Robert Smithson.

It remains difficult to assess the discontinuation of these theoretical attempts in the United States after 1975. One reason might be that many artists returned to the confines of the art system. Another that the Earth Art and Art and Technology movements have always been described as movements with a (more or less) clear beginning and end, leaving no necessity for further evaluation. A third that the interests of art had moved into different directions by now and thus those of the critics and curators. Yet another plausible reason is that the artists, who kept working in the domain of public and media art forms, were outside the mainstream of the official art system.

However, meanwhile voices about the 'death' of the avant-garde also became a little louder. Sociologist Niklas Luhmann put it succinctly when he said that the arts had arrived at a point where development was no longer possible, and that art had reached a point of its own "autopoietische Selbstproduktion" (autopoetic self-production), because art had become purely selfreferential and immanent, operating from within as 'art about art.' Michael Lingner, quoting Luhmann, argues: "Während bisher die Fortsetzbarkeit der Kunst in die Gesellschaftsstruktur eingehängt und dadurch garantiert war, muß die Kunst ihr Überleben allein aus sich selbst heraus sichern, seitdem sie sich wie autopoietische, selbstreferentielle Systeme verhält, welche die Elemente aus denen sie bestehen, durch die Elemente, aus denen sie bestehen zu produzieren haben." (Whilst until now the continuity of art was embedded in the structure of society and thus warranted, art has to secure its survival from within, since it behaves like an autopoetic selfreferential system which has to produce the elements it is composed of by itself.[12]) So, according to Luhmann and Lingner, the visual arts as they exist have reached a dead end. Michael Lingner also concluded that: "Die postmoderne Kunst ist vielmehr als die letzte Stufe konzeptioneller Autonomie zu Begreifen." (Postmodern art can only be seen as the last step of a conceptual autonomy.)[13] In other words, these authors argue that the visual arts are in need of a complete re-orientation; a change of paradigm, to speak with Thomas Kuhn.

The art in the land 'movement' continued and a few specialists in the field, like John Beardsley and Michael Auping, kept writing and organizing exhibitions. The National Endowment for the Arts published an inventory of its art in public places program, for example. A few theoretical attempts were made to situate the larger context of these works. Rosalind Krauss made a structuralist attempt in an *October* article titled "Sculpture in the expanded field." Lucy R. Lippard's book *Overlay* described a much wider array of activities, mostly from a formal (as well as feminist) perspective relating outdoor works to prehistoric forms of art. Mythic and ritual aspects from cultures past also came into play. Again, what we see is an 'overlay' of one or two elements leaving out the idea of the work of art constituting a multiple set of relationships in context as the starting point for the interpretation and meaning of the work as a whole.[14] Recently current public art activities and public art activist activities have been traced back to their roots in the sixties in three publications, setting the stage for a new theoretical approach.[15]

Technology-based art works have a similar history of publications. As multi-media spectacles disappeared from the scene, and increasing sophistication of the advanced technologies almost forced the artist to specialize, video art was soon singled out by the art world as art. As a consequence, its major feature as communications medium was largely overlooked by critics, and so were the initial onsets to discuss the medium in a larger context of media theory or information theory, for example. Here, too, there are a few exceptions, such as

Gene Youngblood's continuing concern with broader social functions of art and artist, as mediator for change. Or, in Germany, Wolfgang Preikschat's vision of a media aesthetics in his *Video Art: Poesie der Neuen Medien.* Meanwhile, the history of computer art forms fell by the wayside until recently.[16] The expansion of digital technologies, followed by the virtual reality and cyberspace inventions, as well as the rapid and unforeseen popularization of world-wide communication networks like Internet, have recently changed this.

Treatment

If one accepts these directions to have been catalysts for a reorientation in the visual arts, it still leaves us with the questions of how and why. For did not Marcel Duchamp already question the precious art object by putting a bicycle wheel and a urinal in the gallery space? What about the rebellious attitude of the Dadaists? And the group ZERO who experimented with random effects and chance already in the fifties? Or the Fluxus actions with their aim to put art on an equal footing with life? Should we not interpret the experiments with natural materials and technological media as a 'natural' evolution frome those movements earlier in the century?

In 1968, Jack Burnham - artist, writer and curator of exhibitions - published a book in which he set out to delineate the relationships between art and technology during the twentieth century as reflected in the change from objet d'art to système d'art. Burnham drew a parallel between the discoveries in the sciences (such as quantum physics) of an invisible conceptual, mathematical world based on interrelated 'open systems,' and the idea that the world was moving toward a systems-oriented structure (communication networks, for example) with the visual arts changing from object to non-object, from material to immaterial, from literal to conceptual.[17] The thesis of *Beyond Modern Sculpture* is not undisputed, yet it remains an important document of the (at times utopian) ideas of those times about the function of a future art within a broader context. Just as important is that it exemplifies that Burnham's considerations came out of, or one might say were part of, a constellation of scientific, theoretical and technological notions which rapidly spread among different disciplines at the time. Jack Burnham's writings and his curatorial activities indisputably portray a section of the puzzle that pictures the events of the late sixties and early seventies. Although Burnham himself soon switched to another theoretical approach, attempting a structuralist analysis of painting, he can be considered a catalyst in promoting the Art and Technology Movement, as theorist and organizer of exhibitions.[18]

In Part I, the developments of Environmental Art forms and the Art and Technology Movement are therefore dealt with practically as well as theoretically in relationship to the new ideas and concepts of systems analysis and cybernetics. This includes a discussion of the organizations which were founded as collaborative ventures with scientists and technicians, allowing artists to familiarize themselves with and obtain access to the current state-of-the-art technological developments and scientific theories and the interdisciplinary activities resulting from this. In Part I also try to explain the filtering down of concepts from systems analysis and cybernetics into different types of art.

Part II deals with artists who moved into the environment, and whose work can be interpreted as a layering of different but related elements. Hans Haacke, who is not now associated with the Earth Art movement, but who began to experiment with natural systems in the early sixties and participated in the first earth art exhibition in 1967, decided to deal with the art system from the inside and made it his life work to critically deal with the museum structure in all its ramifications: economic, social and political, aesthetic, and historical alike. California-based artists Newton Harrison and Robert Irwin left their respective studios after the *Art & Technology Program of the Los Angeles County Museum of Art* in 1970. Although Newton Harrison's initial move resulted from (then) felt restrictions imposed on the designation 'artist,' he soon focused on possibilities for art to function as a problem solving activity, and turned to interdisciplinary research of self-sustaining eco-systems.[19] Robert Irwin, who gave up his Venice (California) studio in 1970, spent the next 2 to 3 years traveling and lecturing to develop his thoughts on the function of art. He even stopped making art objects altogether, saying "To be an artist is not a matter of making paintings or objects at all. What we are really dealing with is our state of consciousness and the shape of our perception."[20]

Artists like Robert Smithson, Hans Haacke, Helen and Newton Harrison, and Robert Irwin as well as Nancy Holt, James Turrell and Alan Sonfist, belong to the 'first generation' who worked in the natural environment with natural systems and who - with the exception of Robert Smithson, who died in an airplane accident - have continued to develop the premises with which they started into a new framework: It will be discussed how this frame of mind includes the human and natural environment as a complex "set of relationships in action," albeit a different context in each case. The starting point became "seeing things in relations."[21]

The chapters in Part III should make clear that the introduction of media like video and the computer, and subsequently other digital technologies, into the visual arts profoundly influenced the modes of production, distribution (presentation) and reception of the work. The works of artists Nam June Paik, Bill Viola, the Vasulkas, and Paul Ryan show the different directions in which artists moved to investigate aspects of these media that are related to the technological functions as well as the social and operational structure. Father of video art, Nam June Paik has often emphasized the communication aspects of video / television. Others, like Bill Viola and Woody and Steina Vasulka have investigated the perceptual and internal structure of video/computer systems. Paul Ryan has from the start sought to integrate his video works in the ecological environmental context. Yet all of them also perceive their work within the larger context of the expanding communication and information networks. These artists have continued to purposively investigate the obligations of the artist in a society like ours, by placing their work and activities outside the studio in the world, as it were, not as a simple object, but in order to set up a relationship with its environment, including the spectator. Moving onto the terrain of other disciplines, inter- and cross-disciplinary efforts and research became almost standard. And sometimes this resulted in a work which was not even recognizable as art.

Starting from the premise that art is a reflection of the society that bears it, it seems legitimate to ask if these developments are in any way related to the increasing complexity of Western society; to ask the question whether the Art in Public Places as well as the electronic Media Arts are not the formal appearance of similar tendencies, requiring an approach or interpretation that takes into account aspects other than only the stylistic or formal qualities. Especially if the artists themselves set out to set up a complex relationship with the public space at large.

The book makes an attempt to contextualize the development of a number of artists, in order to re-assess the function of the art works as well as the position of the artist. It is not intended to construct a well-defined theory, only to put forth some theorems that explicate the interdependency of certain developments in the visual arts and those in an information/knowledge society of "organized complexity" (Daniel Bell). I have purposively selected those artists whose writing sustains this thesis, realizing that I have left out others whose work would also fall within the scope of this book.

I have also chosen to discuss the works of the selected artists as a whole, and not thematically. The reader will notice that certain aspects keep recurring, such as time-space, process, chance or indeterminacy, references to the history of the site, and so on. There are also themes like the maze, or the garden, or geometric patterns like the circle or spiral which are used regularly. However, this would mean that the different aspects would be dealt with as unrelated, whereas they function as interdependent entities. As in chemistry the exchange of one molecule will cause a substance to transform into another, so it is with these art works.

Chapter 1

Interdisciplinary Collaborations: Ideas

Art and Technology Movement

The twentieth century is characterized by several waves of dialogues between art and science and art and technology.[22] In general, the scientific and technical interests of the artists followed those in the sciences and technological application with a time-lag of twenty years. The sculptors of the 1920s, for example, were influenced by scientific models precisely at the time when models (for example the atomic model) were beginning to lose their importance for the mathematician and physicist, who were by then working with the field model and pure mathematics. In the 1950s and 1960s those kinetic art forms, focusing on the perception of movement in light and color images, still relied on mechanical instead of electronic technologies, with the exception of light programming. Also the repetitive field structure of many optical works was in fact based on an already obsolete scientific field concept.[23]

Marshall McLuhan has stated that a particular technology becomes a subject for the artist only when it is already superseded by the next one. Pontus Hultén exemplified this somewhat in his catalogue essay for *The Machine as Seen at the End of the Mechanical Age*, an exhibition about the use of the machine in the visual arts, organized for the Museum of Modern Art in New York in 1968. Be that as it may, these discussions still made a distinction between the relationship between art and science and that between art and technology. It was not until the sixties that this discrepancy in time between scientific discovery and technological application began to disappear. Science and technology have become closely interconnected since then, for both economic as well as political reasons. As a consequence, relationships of art to science and technology were predominantly discussed in one context, that of technology. Simultaneously, there was the introduction of the electronic media and the computer technologies, making their first impact; media that have infiltrated our daily existence. With it came theoretical and philosophical publications about the possible social, political and economic effects of the new technologies, some of which were widely read in art circles.

Literary and philosophical discussions since modern industrialization, i.e. the Industrial Revolution, have often stressed the growing chasm between a scientific culture, founded on specialization of external knowledge, and a literary-artistic culture aiming at intellectual internalization of cultural achievements.Writers, like C.P. Snow, have often perceived of the two as "two cultures," whereby anxious thoughts about a forthcoming alienation were voiced as a consequence of the growing importance of the scientific and technological culture in the society of the Machine Age. For what could be the function of the arts in an automated society? Discussions about possible congruences and differences between the two worlds also predominated in the sixties. Virtually only Susan Sontag put forth a different point of view in her article "One Culture and the New Sensibility" (1966), criticizing Snow's "false assumption that science and technology are in motion, while the arts are static." She perceived an overall new kind of sensibility, which will affect the "transformation of the function of art." Sontag argued that science and technology, all forms of social life, all cultural forms, including works of art, reflect each other, therefore change with each other. For Sontag it followed logically that today "its spirit of exactness, its sense of 'research' and 'problems,' is closer to the spirit of science than of art in the old-fashioned sense."[24] The experimentation with new materials, techniques and forms was its first expression and in a sense this book extended these thoughts.

The wave went into the history books as the Art and Technology Movement. The museum and gallery world reacted with a number of large exhibitions. To mention just the most important ones: Pontus Hultén's *The Machine as Seen at the End of the Mechanical Age*, at the Museum of Modern Art in New York, 1968, presented a diverse survey of machine-related art forms. Hultén had invited Experiments in Art and Technology to participate in the contemporary section. Jasia Reichardt organized a show of computer-based art forms, titled *Cybernetic Serendipity*, which took place at the Institute of Contemporary Art in London in 1970, and traveled to Washington D.C. Then there was Jack Burnham's notorious *Software* exhibition at the Jewish Museum, also in New York, in 1970. Maurice Tuchman initiated a collaborative project to connect artists with industry for the Los Angeles County Museum of Art, the *Art and Technology Program* (1969-1971), and György Kepes organized *Explorations* at the Smithsonian Institution of the National Gallery of Art in Washington D.C., in

1970. There were numerous activities by small interdisciplinary groups like Pulsa and USCO, which discontinued their activities long ago. Pulsa was a group of seven artists and architects teaching at Yale University, who did a series of outdoor temporal light environments between 1968 and 1971, revolving around the concept of the city as artwork, using what they called its "soft information systems," such as water supply, electricity, telephone, heating, as working parameters. USCO was a group of artists and engineers pioneering with multi-media and kinesthetic performances in the United States as well as Europe. In the late sixties they also worked with behavioral scientists "to explore multi-channel techniques and design of facilities, hardware and software." Their ultimate focus was more toward entertainment and as education.[25]

Before discussing interdisciplinary collaborations as the most important characteristic of the art and technology movement, it needs to be emphasized that it was never really a movement in the formal or stylistic sense, nor was it a movement with a uniform set of ideas or concepts. Rather it was the umbrella for a change in attitude toward technology and science among a group of artists. This 'group' of artists was part of a larger 'group' that wanted to extend the boundaries of art. The interest in technology was from the start related to the desire for something new outside the realm of art; a desire to participate in the possibilities for radical changes in our environment which technology was to bring about. Around 1965/66 there existed a situation in which artists interested in technology were looking for knowledge and access. For in discussing the relationship between art and technology, one major problem was the access to this technology. At first, a characteristic of technological progress was the widening of the gap between the sophisticated technologies used in scientific laboratories and those available to the general public. Artists were not able to use these technologies until the personal computer and its accompanying software programs became gradually accessible, and not until the introduction of equipment like the portable video camera and recorder by Sony Corporation, Japan. In addition, sufficient knowledge was wanted. Art academies, colleges and universities were at first not at all equipped to deal with these rapid changes. In reviewing publications, newspapers and magazines, it becomes apparent that the atmosphere was one of an incredibly optimistic belief in the advancement of technology. It stood for progress. The same optimism was voiced in art criticism. John Perrault, critic for the *Village Voice*, wrote: "Certain similarities between art and technology indicate the possibility of a fruitful marriage. These similarities also clarify differences. Art and technology are both systems of translation. Technology translates pure science from the realm of abstract ideas into useful techniques and devices. Art translates experience and emotions into form. ... We are in a period of transition. With one foot in the Mechanical Age and the other in the Electronic, there is always the possibility to slip back into the dark. The Electronic Age, not yet completely assimilated, already seems to be branching off into the Bio-Chemical. Where are we going? ... Our relationships with machines and electronic systems are mirror images of our relationships with ourselves. If we learn how to relate to technology, we may yet learn how to relate harmoniously to ourselves."[26] His writing is an example of the extent to which this advancement was personalized as a utopian possibility for total personal harmony, far beyond what Marshall McLuhan advocated as being a reflection of society's so-called harmonious progress.

Catalysts: György Kepes, Billy Klüver and Jack Burnham

On the one hand, artists sought to find a new function for their work by going outdoors and using the natural environment itself as material; on the other hand, artists sought a new role by exploring the new electronic technologies that were rapidly introduced. In retrospect, two important figures who functioned as catalysts in proclaiming a new relationship between artists and scientists and technologists were György Kepes and Billy Klüver. György Kepes worked since 1946 as Professor of Visual Design at the Massachusetts Institute of Technology, Cambridge.[27] Billy Klüver was an electrical engineer who worked at Bell Telephone Laboratories in Murray Hill, NJ, since 1959. The role of theorist was taken up by Jack Burnham, who promoted his ideas in writings and exhibitions. Kepes and Klüver set out to promote and facilitate collaborative and interdisciplinary projects between artist, scientists, engineers and technicians: Klüver by creating Experiments in Art and Technology in New York, Kepes by founding the Center for Advanced Visual Studies at the Massachusetts Institute of Technology, Cambridge.

Hungarian-born György Kepes had become acquainted with the theories of the Bauhaus and Constructivist movements as well as

Dadaist expressions while moving in progressive circles in Budapest. He was apparently very impressed with the Dada photomontages of Hanna Hoech and Raoul Hausmann. During the late 1920s and early 1930s he changed from photography and photocollage to making films. In the meantime Kepes had become familiar with the work of Laszlo Moholy-Nagy and collaborated with him in Berlin and London between 1930-1937. During his stay in London he also made the acquaintance of a science writer, J.J. Crowther, and it was through him that Kepes met some of England's leading scientists - Joseph Needham, J.D. Bernal, and C.H. Waddington. When Moholy-Nagy invited him to Chicago in 1937 to head the Light and Color Department of the Institute of Design (also called the New Bauhaus) in order to "form a nucleus for an independent reliable educational center where art, science, technology will be united into a creative program," Kepes accepted. The intentions of the program were just in line with ideas that were gradually developing in Kepes's mind; ideas that connected his thoughts on visual communication with the Bauhaus point of view, which attempted to find a philosophical and functional agreement across a wide spectrum of disciplines.[28] There he developed light and color workshops in which various forms and techniques were researched on their visual and psychological impact. However, in 1945 he was invited to introduce a series of visual design courses in the School of Architecture and Planning, at the Massachusetts Institute of Technology, Cambridge (MIT), something new at the time, and was subsequently offered the position of Professor of Visual Design. His association with MIT would prove to be a long and fruitful one. During the following decades György Kepes put his ideas on paper in a number of books which became widely publicized and gave him wide recognition as educator and theorist.

The Language of Vision (1944) was his first attempt to connect the different languages of the different disciplines of knowledge, taking visual communication as a starting point. In particular the idioms of contemporary media like photography, motion pictures and television functioned in his eyes as a universal language that could be understood internationally.[29] Coming from the Bauhaus and Constructivism, György Kepes was of course highly interested in an integration of the visual arts with the visual idioms of the daily environment, including design and architecture. He was able to develop this ideal vision through the interdisciplinary research groups which germinated at MIT around that time. This development was partly accelerated because of the school's involvement in war research as a consequence of the United States' activities in World War II. In 1942, a secret research group came together under the title Radar Laboratory. It was their task to develop a new electromagnetic radar system or anti-ballistic missile system, called Radio Detecting and Ranging. Rad Lab's success was succeeded by the first interdisciplinary (a new term) laboratory at MIT: the Research Laboratory of Electronics in 1952. Its director was Jerome Wiesner, and one of the participants was Norbert Wiener, who was making a name in thc science of cybernetics at that time. Wiesner in an interview with Steward Brand: "The Research Laboratory of Electronics was probably the most exciting place in the world for anyone interested in communications. We were doing research on neurophysiology, we were studying electrical noise problems, we were doing coding, we were following Shannon's work on information theory, we were even thinking about computers. Out of this I acquired the idea from Norbert Wiener that we would understand both living system communications and machines better if we worked on them not necessarily together but in the same environment."[30]

It should come as no surprise that the interdisciplinary nature of these projects triggered Kepes's interest to research possible relationships between art and science. While there was a growing exchange of ideas amongst a number of science departments, he noticed very little discourse between the humanities and science faculties. But even if many people thought that art and science were unmixable entities, Kepes was convinced that there existed a symbiotic relationship between the two, which would only grow stronger when nourished by each other.[31] Informed about the "new frontiers of knowledge," as he called the scientific discoveries, and aware of the potential of technological progress, he expressed both his admiration and ambivalence. "Science appears to be the most confident and vigorous unifying force today. It has produced a new foundation for our material existence, but as we all know, has not yet solved our human problems. So far, our recently acquired knowledge, with all its precision and power, has brought us as much ugliness, discomfort and danger as it has sanity and order. Science is only one component of the understanding that we need for a well-balanced attainment of human ends. In our chaotic and directionless world, it gives us two-edged weapons - powerful tools and ideas with which we may either create or destroy."[32] For Kepes, technological progress was not in itself negative, but he warned against an uncontrolled utilization of it. György Kepes: "The obvious world that we know on gross levels of sight, sound taste and touch, can be connected with the subtle world revealed by our scientific instruments and devices. Seen together, aerial maps of river estuaries and road systems, feathers, fern leaves, branching blood vessels, nerve ganglia, electron micrographs of crystals and the tree-like patterns of electrical discharge-figures are connected, although they are vastly different in place, origin and scale ... Their similarity of form is by no means accidental. As patterns of energy-gathering and energy-distribution, they are similar graphs generated by similar processes."[33]

His interest in the progress of science and technology was always guided by an ethical attitude. But he went even further, saying that we should never lose sight of its effects on the environment, and that the technical landscape should in fact serve to restore a lost balance of nature. When Kepes tried to give an account of the different functions of the visual arts and the sciences, and what might connect them, he kept returning to the idea that nature served as a common base, as a common language for both. Being both ordering devices in that they try to impose a structure upon the human experience, art and science could have a complementary function in restoring the balance in today's society, which he felt had been lost. What connected them further, argued Kepes, was that both art and science were "image-making devices" which needed to visualize experiences. For György Kepes it was no accident that there was such a remarkable similarity between certain paintings and photographs and the images that had become visible through new optical technologies such as infrared and ultraviolet rays, microscopic and telescopic photography, X-rays and other radiation techniques. Both were looking for laws, such as pattern, structure, harmony, order or even disorder in natural and other phenomena. These technologies offered a completely new picture of nature's order, which had hitherto been invisible to the human eye, giving new sensory experiences and expanding the range of perception. In order to make his findings known to the public, he organized an exhibition titled *The New Landscape* (1951), in which he presented the visual analogues between recent scientific visualizations of research models and the visual arts. Computer images, serial and other photographs of the "micro-world" and "macro-world" made by scientists, were presented next to artist's images, showing analogous forms and patterns. A few years later he published *The New Landscape in Art and Science* (1956), in which he "attempted to present in pictures the new visual world revealed by science and technology, things that were previously too big or too small, too opaque or too fast for the unaided eye to see."[34] This book remained relatively unnoticed for a while. This was because his timing was off at the time, he thought later. Many scientists and artists still considered art and science unmixable entities.[35]

György Kepes gradually moved beyond a comparison of visual analogues, on to another level of relationships between art and science. He felt that the scientific concerns with the invisible micro-world in terms of energy or dynamic organization instead of measurable, and tangible objects could be compared to the changes in the visual arts. The fact that sculptors had turned to new 'immaterial' technologies of light, video or laser was representative of this new way of thinking for him. His position allowed him to closely follow the latest scientific developments, which not long before had turned the classical mechanistic notions of the universe upside down, absorbing the new ideas about time and space, about seeing things as an interrelated whole, and about the fusion of the subject-object distinction. He introduced these concepts in his writing to give a theoretical basis to the visual analogues he saw in art and scientific imagery.[36] He had also come to understand that the scientists were looking for new ways to describe and visualize their 'abstract' experiments which were making science as a whole look more and more like a conceptual construct of mathematical equations. Kepes was convinced that if artists and scientists were to work in a closer communion, it might be possible that artists would work out new visual images that could inspire scientists in their search for new visual models.

Billy (J.William) Klüver was born in Monaco in 1927. He studied at and graduated from the Royal Institute of Technology in Stockholm, Sweden in 1951, and moved to the United States in 1954 to obtain a Ph.D. in Electrical Engineering from the University of California at Berkeley. From 1958 to 1968 he was on the Technical Staff at Bell Telephone Laboratories, in Murray Hill, New Jersey, as a research engineer specializing in laser technology. But his interests went beyond the research laboratory. Since the early fifties he maintained a friendship with Jean Tinguely, whom he had assisted with his motorized devices, and with Pontus Hultén, director of the Moderna Museet in

Stockholm then. Hultén was very interested in artistic phenomena that revolved around motion or movement in art and had organized an exhibition *Art in Motion* in 1961, for which Klüver had put together the American contribution. After the famous event *Homage to New York* (1960), for which Klüver had assisted Jean Tinguely in building a machine that would destroy itself in the Museum of Modern Art's garden, he extended his activities to work with other artists during the following six years. He worked on Merce Cunningham's *Variations V*, Warhol's helium-filled 'clouds,' Jasper Johns' paintings with neon lights, Claes Oldenburg's *Store Days*, Robert Rauschenberg's *Oracle* (1965) and *Soundings* (1968). *Soundings* consists of nine 8-foot-high panels with silvered imagery (silk-screen) on plexiglass, in which the viewer at first only sees his own reflection. When the viewer speaks, the work responds to the varying resonant frequencies of a speaker's voice. For different combinations of words and sounds, different combinations of chairs appear and disappear. Robert Rauschenberg worked with a team of engineers from Bell Laboratories, who designed and wired up the panel for 200 combinations of response. For a while the Bell Laboratories seem indeed to have been a central meeting point for artists and engineers. Klüver: "From 1960 on I brought some 50 to 100 artists through Bell Labs. I don't know how many. I had a tour to show them what went on, and this mounted an incredible interest in technology. You would read in the newspapers and everywhere else about the advancement in technology. Nobody could have access to it. So the fact that I took them on these picnics out at Bell Labs obviously was memorable. Marcel Duchamp came there. Everybody came there. And that did probably more than anything. Because the artists could see the hands-on process and that one could actually DO things."[37]

One can imagine that this atmosphere was particularly fruitful for the circle of New York artists which Klüver frequented and vice versa, sowing the seeds for new relationships between artists, technologists and scientists. He was very interested in art and in what effect the artist could have on technology. Although the scientist's vision and realization differed widely from that of the artist and engineer in his opinion, because the artist and the engineer were dealing with the physical world, and worked directly toward the solution of a problem, contrary to the scientist, who was trained to understand it, there was never any question about a scientist becoming an artist, or an artist becoming a scientist. There was no doubt in his mind that the artist needed the engineer, because of the growing complexities of technology.

His theoretical statements, whether aesthetic, social or political, were controversial at times. For example, he declared that it makes no difference whether in a purely technocratic society technology really functions in art, in opposition to the machine which must always be a functional object. He still defined art as purely non-functional, a notion that suited the work of Jean Tinguely. Tinguely's machines are totally dysfunctional indeed. As a collaborator Klüver was basically interested in providing the scientific and technical unit that would make the artist's idea work. "The function of technology as a material is not to put previous aesthetic concepts into new forms but to provide the basis for a new aesthetic, one that has an organic relationship with the contemporary world," said Klüver in an interview with Douglas Davis.[38]

New Relationships between Art, Technology and Science

After the publication of *The New Landscape in Art and Science*, György Kepes's contributions to books and catalogues almost always revolved around a serious concern about the earth and the environment, emphasizing the thoughtless manner in which mankind treats nature's treasures. His arguments grew stronger in the early seventies when the ecological movement was rapidly gaining a large following and environmental concerns were vehemently and emotionally discussed in books like Charles Reich's *The Greening of America* or Rachel Carson's *Silent Spring*. Kepes: "Our cities are our collective self-portraits, images of hollowness and chaos. And if not properly guided our immensely potent technology may carry within itself curses of even more awesome proportions. The yet not understood uncontrolled dynamics of scientific technology could do more than poison our earth; it is capable of wreaking havoc on man's genetic nature."[39] It is here that Kepes found a social as well as educational role for the artist. Only an artist who was knowledgeable about the developments in science and understood the implications of new technologies could influence the expected environmental problems. For Kepes, this meant a new "civic art" (his term), based on interdisciplinary collaboration. One way to expand the vision of art was to create an exchange and communication for art with other disciplines, in particular the sciences. "In becoming a collaborative enterprise in which artists, scientists, urban planners and engineers are interdependent, art clearly enters a new phase of orientation in which its prime goal is the revitalization of the entire human environment; a greatly-to-be-wished-for climax to the rebuilding of our present urban world."[40]

At the time Kepes's ideas about a civic art appeared quite new in American eyes, because they stood in sharp contrast to the prevailing situation in public sculpture. A brief look reveals that public art was simply defined as an art object located in a public place. It was generally identified with monumental objects, historical personalities, spouting fountain nymphs, or playful animals, that fill in an open spot in a park, a plaza or a bridge. Even the National Endowment for the Arts, founded in 1965, whose original aim was to promote more experimental endeavors in art, initially commissioned well-known artists to create an object-sculpture, and subsequently donated it to a city for siting in a public area. One can hardly blame the NEA for taking this well-intended position, though, since there were hardly any other public art programs in the country at the time.[41] Thus, when György Kepes laid down his ideas toward a civic art, he had something different in mind than these traditional monumental objects which embellished the parks and plazas of so many towns. His suggestions of "water purification kinetic sculptures, which could be framed with water gardens or water parks, providing an intricate interplay of all varieties of water movements and sprays in different densities and shapes," or "immense transparent structures that give visibility to hydraulic processes, a contained but legible ballet of water racing through obstacles of filters, tinted and purified by chemicals, or moving sluggishly in intricate but legible patterns of transparent containers," may not even sound strange in our ears. However, they were fantastic and rather idealistic, and the realization appears somewhat naive now.[42]

Although Kepes showed great foresight concerning the environmental problems that were looming ahead, the solutions he offered were either too general and vague or too comprehensive. Even if his ideas appeared a little vague and utopian, a problem which was enhanced by a language that lacked a more precise definition of his civic art, it is to Kepes's credit that he saw that the existing public art forms had reached a dead end. He was one of the first artists who understood that a new direction was needed, and that this direction had to go together with a new function for the visual arts and artists. It should be noted, though, that György Kepes's visions about a harmoniously functioning human ecology, which he thought could be achieved by resolving the major disturbances of both the individual as well as communal aspects of our lives, were similar to the messages in contemporary anthropology, psychology, or the applied sciences. For anthropologists had begun to study 'primitive man' as part of all-embracing social, cultural and natural cosmic surroundings. The applied sciences had started to present thought models of dynamic interconnectedness of seemingly disparate processes and systems, particularly in fields such as computer technology, electronics and communication networks. Simultaneously, Marshall Mcluhan ventilated his utopian idcas about the world as a "global village," functioning harmoniously with the "tribal drums" of electric (electronic) communications. The Club of Rome, founded in 1968, set out to create an understanding for the interdependence of political, social, economic, scientific, technological, ecological and cultural aspects of our current worldwide problems, introducing a systems approach as a method to get a handle on what they considered the future scenario of the world. This is also its main merit. The Club introduced a global approach, which was in itself positive but not purposeful. Meanwhile its strategy has evolved into a less comprehensive one, which does not propose to solve the totality of these problems, but deals with certain issues separately as one aspect of a set of related problems.[43] Yet in the sixties meteorologists still thought they could change and control the climates of the earth, and economic scientists argued that they could control national economic situations. Only recently scientists and politicians seem to understand the significance of the enormous task before us somewhat better, and propose solutions and treatments for single issues, which are now treated as a complex of interacting forces or actions.

As a professional engineer working in a research laboratory for communication technologies, Billy Klüver understood directly the consequences that a rapid spread of technical information might have, and its possible influence on the human environment. His fear was that art, the art-intellectual world, would move further and further away from technology, and get more and more frustrated in not being able to handle its complexities. In 1967 Billy Klüver founded Experiments in Art and Technology (E.A.T.) together with Robert Rauschenberg, convinced that a working relationship between engineers, technicians and artists, in which both would have an equal share in the creative process, would further the understanding of each other's worlds, and in so doing shape a relationship that would create possibilities which would benefit society as a whole. If the artist were to learn about the possibilities of technology, and the engineer were stimulated by the ideas of the artist, both might learn to understand each other's language. He hoped that this might conceivably lead to an increased awareness of the current positions and processes of the artist and engineer respectively, and ultimately to a renewed reflection on art's and technology's function in society. He envisioned that it might even lead to a true concern with the larger, environmental issues of the earth.[44] Klüver speculated that the main future influence of art and technology could be in the area of environmental issues. He hoped that by using technology as material the artist could develop a new aesthetic that would have an organic relationship with the contemporary world. What this in fact implied, was a new definition of the artist's role in society. For no one could defy the environmental consequences of technological change any longer, he wrote.[45] Both Kepes and Klüver felt that artists had to take responsibility in a social restructuring of society by educating the public through a new visual aesthetics, and in order to do so they would have to leave the studio, reorient themselves and enter into a dialogue with the real world. The prevailing idealistic concept underlying these ideas was that artists were indeed the group of people who were the first ones to detect necessities for change and react. They were guided by the ideal that art could really influence the developments to come. Klüver's consideration was that, by participating creatively in industrial and communal processes, they could exert some power and possibly influence decision-making and things to come. For him, this meant not so much that the traditional function of art was totally abandoned, but that art would play a truly functional role in the changing of societal structures.[46]

In contrast, Jack Burnham's vision moved beyond the existing, when he envisioned a function of art in what he called a "post-biological logic for technological development," if "survival, adaptation, and regeneration form the cornerstones of biological existence, it may be that culture is fundamentally a means for implementing qualitative transformations of man's biological status." Burnham continued: "Art, then, and the whole image-making drive may be a means for preparing man for physical and mental changes which he will in time make upon himself. Sculpture, functioning so, becomes a kind of physical radar signal preparing the human race thousands (or now perhaps scores) of years in advance"[47] Burnham's teleology already envisioned a world living by biochemical engineering and genetics, i.e. a world in which the traditional art object or sculpture would have lost its function. It was clear to Burnham that the visual arts were beginning to adapt to a new situation indeed, when one set out to draw parallels between the developments in the sciences and technology and the visual arts, as he did in his book, *Beyond Modern Sculpture* (1968). His view was that art had already changed from being an "objet d'art" to a "système d'art," which implied that the new art work or sculpture was characterized by a set of relationships, and no longer defined by its boundaries, according to the concepts of the general system theory and cybernetics, which were rapidly gaining popularity. He found his theory confirmed in the work of Hans Haacke, Robert Morris and Dennis Oppenheim, who were no longer making objects with a well-defined frame, but chose to make natural processes or actions in the environment the content of the art work.

He went as far as to state that the new technologies might lead to a dissolution of art as we define it, and he tried to demonstrate this idea organizing an exhibition, *Software* (1970), to show the different creative possibilities of computer-based art. In the introduction to the catalogue, he explained that "*Software* is not specifically a demonstration of engineering know-how, not for that matter an art exhibition. Rather in a limited sense, it demonstrates the effects of contemporary control and communication techniques in the hands of artists... *Software* makes no distinctions between art and non-art... Hence, the goal of *Software* is to focus on... information processing systems and their devices."[48] For Burnham, there was no doubt - then - that the information and communication technologies would soon permeate our daily existence and change our aesthetic perception. Also, Burnham held the opinion that, if artists wanted to have an influence on this process, they would have to research how information is controlled, who has power over and access to it, how information processing would affect the psychological outlook of the average human being on questions of protection of information, of ownership, of privacy, and so forth, in short, how the technology at large functions.[49]

Like Klüver and Kepes, Burnham too stated that it was the task of the artist to take responsibility for the well-being and continued existence of life upon earth. But contrary to the others, he was very specific in his theoretical assumptions that the new function for the artist was to be found in the communications technologies and information networks and toward a functional aesthetic (Benjamin Buchloh's term), as it was developed by, for example, Hans Haacke, in a society dominated by information systems.

Chapter 2

Interdisciplinary Collaborations: Projects

The *Nine Evenings*

Billy Klüver's ideal was that engineers and artists would have an equal share in the creative process from the beginning, so that a viewer would not be able to distinguish the different contributions in the result. With this in mind Billy Klüver and Robert Rauschenberg became involved in the organization of the notorious and famous *Nine Evenings: Theatre and Engineering,* which took place at the Armoury Building in New York City, from October 13 - 23, 1966. Originally, a festival of art and technology had been scheduled in Stockholm for the fall of 1966 where Klüver was asked to participate and assist with technical support. The Swedish festival fell through and the American version turned into the *Nine Evenings.* The idea behind the project was that artists would collaborate with engineers on the project. The engineer would not only furnish technical assistance, but both would be equal partners in the creative process. In his introduction to the catalogue, Klüver said: "... 9 Evenings is a truly cooperative venture. All participants had an equal voice in the direction and all responsibility was shared jointly. The objectives of the 9 Evenings will be continued by Experiments in Art and Technology, Inc. This foundation will further the creative interaction between industry, engineers and artists. *9 Evenings* is an experiment in the true sense of the word: its results are open for the future."[50] Robert Rauschenberg, who had already collaborated with Klüver on a few projects, expressed himself clearly about the use of recent technologies by artists when he said: "Working with engineers is inspiring. I could not do what I want to do without them. It is no longer possible to by-pass the whole area of technology. We have no assurance, for example, that buildings will have walls for much longer. I can foresee art schools giving courses in electronics and vacuum mouldings. We can't afford to wait. We must force a relationship with technology in order to continue and we must move quickly. The most positive thing I can say is that technology does not lead us back into history, but advances us into the unknown."[51]

Among the participants were Robert Rauschenberg himself, John Cage, David Tudor, Robert Whitman, and Lucinda Childs. To create a visual impression of the performances, I have selected a few descriptions from the catalogue, and from newspaper and art reviews. They vary considerably in clarity; some emphasize the visual sequences, others the technological aspects. John Cage wrote about his performance *Variations VII*: "It is a piece of music indeterminate in form and detail... making use of the sound system which has been devised collectively for this festival.. using as sound sources only those sounds which are in the air at the moment of performance, picked up via communication bands, telephone lines, and microphones, together with, instead of musical instruments, a variety of household appliances and frequency generators." David Tudor's *Bandoneon!* "incorporated programmed audio circuits, moving loud-speakers, TV images and lights." Tudor: "..(it) will create signals which are simultaneously used as materials for differentiated audio spectrums (achieved through modulation means, and special loud-speaker construction), for the production of visual images...for the activation of programming devices controlling the audio visual environment."[52] Robert Rauschenberg's contribution to the *Nine Evenings* must have been one of the more spectacular of the evenings. Nilo Lindgren described it as follows: "It takes place in the enormous half-barreled space. The Armoury floor becomes a tennis court, and as the two players hit the ball back and forth, its sound against the racket is transmitted from the radios inside the racket handles, and amplified through speakers. As the play proceeds, the lights illuminating the court are extinguished one by one each time the ball is struck, until the stage is dark, at which the game ends. At that point, hundreds of people move onto the Armoury floor where they are illuminated with infrared lights; the crowd scene in the darkness is picked up with infrared sensitive television cameras, then projected on three large screens, so that the audience feels the presence of a large crowd without being able to see it except on the reproduced screen images. When the lights go finally on, the huge crowd of 'actors' bows to the audience seated on the bleachers."[53] About Whitman, he continued: "In (Robert) Whitman's piece *The Two Holes of Water,* the Armory space, in relative darkness, gradually fills up with automobiles that drive slowly, sheathed in plastic, announced by loud sounds as of warfare and political speeches. The cars maneuver into positions around movie screens that appear one by one around the Armory space, until one is viewing simultaneously about ten screens showing animals running, penguins frolicking, planes taking off and landing backwards, cells dividing, a man and a woman dressing and undressing, caressing, silently mouthed television advertisements, and landscapes going by, accompanied by the sounds of bombs exploding, aircraft, political harangues, typewriters clacking, and so on..."[54],[55]

The performances received a highly critical response. Brian O'Doherty wrote in his critique: "The evenings received, on the whole, an appalling press, based mainly on the justifiable irritation of interminable delays, technical failures of the most basic sort, and long, dead spaces between, and sometimes in the middle of pieces. Yet, as such irritation faded away, one is left with startlingly persistent residual images, and strong hints of an alternative theater that has been lagging in its post-Happenings penumbra between art and theater..."[56] Jack Burnham on the contrary reviewed: "Few if any had the prescience to appreciate the events for what they were: man-made systems with a completely different set of values from those found in structured dramatics or the one-night kinetic spectacular. ... This suggests that systems-oriented art - dropping the term 'sculpture' - will deal less with artifacts contrived from their formal value, and increasingly with *men* enmeshed with and within purposeful responsive systems. Such a change should gradually diminish the distinction between biological and non-biological systems, i.e. man and the system as similarly functioning but organizationally separate entities. The outcome will neither be the fragile cybernetic organisms now built nor the cumbersome electronic environments just coming into being. Rather, the system itself will be made intelligent and sensitive to the human invading its territorial and sensorial domain."[57] His review moved far beyond the works itself into speculation on future possibilities, which have partially become true, though not as far as the official art world is concerned.

It is significant that the engineers' contributions were seldom mentioned in art critiques. However, without their contributions the performances would never have turned out the same. The only one who extensively described the mutual contributions was Nilo Lindgren, staff writer for *IEEE Spectrum*, a magazine for the engineers' community. He described in detail the initial development process behind the performances, as well as the roles of Billy Klüver and Robert Rauschenberg as the two major driving forces in the collaboration process. He was also the only one who presented an analysis of both the artist's and engineer's language problems in communicating. He asked himself the question "what is in it for the engineer?," a question not generally put forth by the participating artists. It indicates that the collaborations were not as easy as Klüver would have us believe. Although his final conclusion was positive, Lindgren: "This new situation, perhaps an uncomfortable and anxiety-making one for both artists and engineers, was the force, the acceptance of new responsibilities - the need to state requirements and concepts clearly and to stick by the engineer as he pursued his design and construction, the obligation to wait while engineering work was being completed, and so on. The very mode of collaboration attacks the old-fashioned idea that the artist is a very special kind of person, in the sense that he is mystical, unmatchable, or incommunicado on his pedestal."[58],[59] The engineers' contributions were remarkable indeed, and not just limited to technical performance. The engineering crew, the majority of which came from Bell Telephone Laboratories, where Billy Klüver worked, designed a.o. a ground effect machine designed by Per Biorn and an 80-kHz Doppler sonar built by Peter Hirsch for Lucinda Child's *Vehicle*; an anti-missile missile and floating snowflakes for Öyvind Fahlstrom's *Kisses Sweeter Than Wine* were designed and built by Harold Hodges, a laser research specialist; Cecil Cooker worked on synthetic computers; Jim McGee, who worked on holograms, built programming drums for the performances; Leonard Robinson designed the power amplifier for a theater electronic environment modulator; Herb Schneider took over performance planning for all the artists; Fred Waldhauer, who worked in communication systems research, designed a proportional control system; Dick Wolff, who did superconductivity research, developed a radio loop system for Paxton's *Physical Things*. Incidentally, the artists and engineers working collaboratively on the *Nine Evenings* made a few discoveries which turned out to be useful for further scientific research or commercially applicable in the near future. They found a phosphorus compound that could be used in infrared laser research, designed a small power amplifier, and a proportional control system, which varied sound and light on stage simply by passing a small flashlight over glass covering sixteen photocells.[60]

Judged by conventional art-historical standards, the *Nine Evenings* belonged to the intermedia works that were flooding the New York art scene at the time. They were theatrical, temporary, and all the performances used a combination of materials, ranging from multiple slide and film projection, light and amplified sound, dance and music, to huge crowds on the stage itself. Obviously the technical features were dominant, but that they could possibly have a decisive influence on the aesthetics of the works was hardly noticed. Apparently the technical failures made a greater impact. The effects of built-in chance processes and randomness were not appreciated. After all, it was only the final result that counted, said Lucy Lippard.[61] Jack Burnham was the only art writer who perceived that these works originated from a different "set of values" and thus should be evaluated differently. Burnham had arrived at the conclusion that use of new materials and new standards of fabrication lead to the development of a new aesthetic which was synchronized with scientific and technical goals. Having also become interested in artificial intelligence, he even went much further, speculating that future intelligent systems would react and adjust to human intervention.

The project was not only the first public event in which technological features were used experimentally to such an extent, but for Douglas Davis they were the result of a new type of interdisciplinary collaboration.[62] Altogether one may conclude that the emphasis on the artist and the engineer as equal partners in the creative process was significant. On the one hand, it implied a change in attitude by the participating engineers and scientists toward the arts, and an attempt to understand the different objectives of art. On the other hand, it required the artist to see the process as team effort and leave as it were some of the quest for authenticity behind. The latter marks the essence of Klüver's ideal, an ideal which, when consistently pursued, also implied a redefinition of the traditional role models of artist and engineer, as well as of the art work. Even if this was not clear at the time, it completely changed the production process from a solitary act to a team effort. This process continued in the presentation and made the viewer a participant in the creation of the work. It is these aspects that were to become important.

While György Kepes's visions focused on the thought that the scientific and the technical 'landscape' must be brought into harmony with the natural environment again, and that the future role of the artists was to further this task by developing large-scale projects that would involve the environment at large, to restore this "lost pageantry in nature," he began to look for practical ways that might initiate such a process. One road was to restore communication between the two totally different worlds of the scientist and the artist. In 1965 Kepes presented and published a proposal for a center that would a a basis for "the formation of a closely knit community of eight to ten promising young artists and designers... The group, located in an academic institution with a strong scientific tradition, would include painters, sculptors, filmmakers, photographers, stage designers, illumination engineers and graphic designers... It is assumed that close and continuous work contact with one another and with the academic community of architects, city planners and engineers would lead to...the development of new ideas."[63] It became the basis for the Center for Advanced Visual Studies at the Massachusetts Institute of Technology founded in 1967.

Billy Klüver felt that the social responsibility of the artist was to look out into the world, into physical reality, without prejudice. This involvement with the physical world logically had to include an interest in technology, technology being, as I said previously, the governing agent in the changes that were taking place in Western society. This involvement with the physical world also meant that the artist would be involved in so called "non-art areas as well as in situations where technology was used to create works of art or was integrated into theater performances," away from the isolated art object.[64] The result was Experiments in Art and Technology, whose practical aim was to bring artists and engineers together to collaborate on an art project. In a recent interview Klüver responded to the question why this all happened at that particular moment: "We were the only ones who took on technology and the artist. Our statements on the purpose of E.A.T. were incredibly utopian. There wasn't any other way to go except that direction. And there was no compromise. We believed that art made a difference."[65]

Experiments in Art and Technology

Whatever the critic's or public's response to the *Nine Evenings,* Billy Klüver found a companion in Robert Rauschenberg. Both felt they had a task to fulfill and on Wednesday October 11, 1967 the official birth of Experiments in Art and Technology (E.A.T.) took place in Rauschenberg's loft. It was announced in the New York Times as "Art and Science Proclaim Alliance in Avant-Garde Loft." The press conference held at Robert Rauschenberg's included films, an exhibition of works incorporating technology and a formal presentation that explained the wide range of Experiments in Art and Technology's program. Still a modest and uncertain merger, it sought to bridge the gap between the two worlds of art and technology. It intended to bring modern technological tools to the artist to facilitate experimentation with new art forms on the one hand, and bring fresh insights and viewpoints to the engineer on the other, in order to make technology more user-friendly as one would say today. The idea of humanizing technology was filtering down in particular among those interested in cybernetics since the publication of Norbert Wiener's popular book *The Human Use of Human Beings: Cybernetics and Society* (1950).

During the preceding year, E.A.T. had set up offices in a loft at 9 East 16th Street, which served as office space, storage for equipment and as a meeting place for artists and engineers. A newsletter was published January 15, 1967, followed by a second one in June, which included a statement of the organization's objectives: "The purpose of Experiments in Art and Technology, Inc. is to catalyze the inevitable active involvement of industry, technology and the arts. E.A.T. has assumed the responsibility of developing an effective relationship between artists and engineers." "E.A.T. will guide the artist in achieving new art through new technology and work for the professional recognition of the engineer's technical contribution within the engineering community." "To ensure a continued fruitful interaction between a rapidly advancing technology and the arts, E.A.T. will work for a high standard technical innovation in collaborative projects." "E.A.T. is founded on the strong belief that an industrially sponsored, effective working relationship between the artists and engineers will lead to new possibilities which will benefit society as a whole."[66] Through contacts with I.E.E.E. (Institute of Electrical and Electronic Engineers), news about E.A.T. soon circulated in the technical press. The obvious enthusiasm and pertinacity with which the staff at E.A.T. pursued its goals contributed greatly to what were considered successful early years.

Obviously E.A.T. provided just those things that artists were looking for: access to technology as well as technical assistance. The financial records show that the objectives combined idealism with a healthy dose of pragmatism. In 1968 already, the organization managed to obtain considerable support from the industrial and private sector, in addition to public grants from Rockefeller Brothers Fund ($25,000) and from the National Endowment for the Arts ($50,000).[67] This tendency continued over the next few years, when the organization was able to obtain support from large industries and commercial organizations as well. After all, the foundation described its long-term policy as: "E.A.T.'s long-term policy in relation to industry is to evolve a permanent, functional relationship between industry and the individual artist in which the industry fully recognizes the contemporary artist as a resource. In developing this relationship, E.A.T. will work for a fair remuneration to the artist for his work and encourage industry not to be embarrassed to recognize the commercial potentials of artists' ideas."

It is no accident that the organization looked for and found such prominent figures as Robert O. Anderson from the Atlantic Richfield Corporation (ARCO), Max Mathews from Bell Telephone Laboratories, or the former U.S. Senator Jacob Javits, to sit on the Board of Directors and Council of Agents, next to György Kepes, Philip Johnson and John Cage.[68] These connections secured at least in part the organization's in-kind and financial support that continues until this day, an achievement that has largely been overlooked. Corporate support was not always uncontested, as reactions to the *Art and Technology Program* of the Los Angeles County Museum of Art. This program placed artists in residence with industries and corporations for a limited period of time to work on a collaborative art project. The *Art and Technology Program* was initiated by Maurice Tuchman in 1967, "to bring together the incredible resources and advanced technology of industry with the equally incredible imagination of the best artists at work today." Artists invited to enter a proposal were put in touch with a company to develop the project during a ten to twelve week artist-in-residency at the plant. Again the idea was to bring artists in touch with new technologies and the industry to induce future collaborations. Eventually, thirty-seven corporations in Southern California, most of them involved with aeronautics and space technology, contributed to the program. Twenty-seven artists worked directly with specific companies and about twenty collaborations resulted in a tangible artwork. Among those was a small group of artists whose research led to the use of advanced technologies at the time, such as lasers, advanced mirror optic systems or gas plasmas.

The majority of the artworks were of a non-object, environmental type, with an emphasis on transient images and evanescent phenomena. A number of the projects were exhibited at the U.S. Pavilion of the

World's Fair in Osaka, Japan, 1970. Critics attacked the large sums of money it had cost, calling it the multimillion dollar art boondoggle.[69] Yet, quite a few industries have opened their doors to artists since then, recognizing the effect that creative input might have on the direction of a research program or project. But that this kind of support was not only a gratuitous gesture, and that these sponsorships might also serve to enhance the reputation of a company culturally, was recognized as well. Corporate funding has become the major source of income for museums and other art institutions.

Although E.A.T.'s early years were considered quite successful, there existed initial confusion about the exact nature of the organization. Initially artists would come to its offices expecting to be given information about technical problems and equipment services, or referrals to facilities. Of course, technicians and engineers did not know quite what to expect either. Nevertheless, we read that nearly 500 matches between artists and engineers were created during the first two years.[70] Artists came to E.A.T. basically for practical reasons, but the organizers have always worked from an underlying idealistic consideration. In the previous chapter, we already mentioned Billy Klüver's ideals about a social function for art, and it is these idea(l)s that gradually became a stronger element in Experiments in Art & Technology's policies. "We can no longer claim innocence for the human and social consequences resulting from technological change. We are, in all respects, responsible for the technology that will form our environment tomorrow. This responsibility implies the search for a technology directed towards pleasure and enjoyment. It implies a distinction between functional and ecological aspects of technology," wrote Klüver.[71]

Experiments in Art and Technology's early projects were art projects in the true sense. The first important exhibitions it was asked to participate in were *The Machine as Seen at the End of the Mechanical Age* (MoMA, New York 1968) and *Some More Beginnings* (Brooklyn Museum of Art, Brooklyn 1968/69), followed by the *Pavilion* for the 1970 World's Fair in Osaka, Japan.[72] Gradually, however, the organization extended its field of operation to include field research for television programming, communication and education, cultural programming, and even housing and agriculture in a program called *Projects Outside Art.* For example, at the request of the RCA Laboratories, in 1969 E.A.T. developed a proposal for a research laboratory in entertainment programming, where artists would work collaboratively with research engineers, sociologists, anthropologists, and so forth, to explore new forms of entertainment programming towards the creation of a more effective role for television's social context. Also in 1969, the Nehru Institute for Development in Ahmedabad, India, invited E.A.T. to collaborate in designing and establishing a field research laboratory for developing methods of producing instructional television software for use in rural villages, in preparation for the SITE satellite network.[73] Yet another project was the generation of a system for regional out-of-studio cultural programming as a basis for a public television channel to produce adult instructional material.

After 1972, artists started to accuse E.A.T. of being more directed toward the science and engineering world than the needs of the artist. Robert Rauschenberg, who had meanwhile withdrawn from the board to focus again on his artistic career, however, commented in an interview with Douglas Davis in 1972: "People grow dissatisfied with us because we're involved with the sociology of art, not the esthetics. For years now, we've been in the middle of the conflict involving art, money, power, and ethics. E.A.T. has been resisted from both sides, by the artists and by industry. We knew we were going to have trouble with the latter, because industry is not noted for its conscience...but the artists turned out to be impatient, too, because they misunderstood us to be a kind of library or source for new materials. But E.A.T. was interested in a total and mutual commitment. We didn't want the artist to use an engineer or an industry to merely execute preconceived ideas but to conduct research in which both sides flatten the ego."[74]

As E.A.T. announced its more wide-ranging goals outside and beyond the artistic world, the projects that E.A.T. became involved in - soon after its foundation in fact - were geared more and more to large interdisciplinary ventures whereby artists were asked to contribute their creative insights in a socially relevant and functional way. Of course, for these large-scale interdisciplinary projects, requiring collaboration and teamwork, and where the result was not anything that resembled an art work with an individual signature, the ego and authorship of the individual artist had to be set aside. One can imagine that, for many artists, this position was stretching the boundaries of art too far. Thus, since the organization had come to depend on funding and support mainly from industrial companies, it turned less toward projects suiting the individual artist, who in turn felt neglected. By 1974/75 the excitement about the art and technology connection had died. Little attention was paid to it in the art world and the art press from then on. Frank Popper's *Art - action and participation* (1975) was the last major publication paying attention to the subject for a long time to come: until the nineties, with the rise of virtual reality and cyberspace technologies.

The *Pavilion*

Exemplary among the art projects and exhibitions that the organization became involved in was the *Pavilion.* As appear to have passed it by most of the written histories on this art and technology episode, it seems worthwhile to narrate it. In August 1968 E.A.T. agreed to conduct supervision over the development of an art project during the Expo '70. Pepsi-Cola Corporation was prepared to sponsor this undertaking, and a team of artists and engineers began to develop ideas, which resulted in "a programmed environment of sound and reflected imagery inside the pavilion."[75] The *Pavilion*'s background remains elusive even if one compares the histories of Nilo Lindgren and Calvin Tomkins.

As is often the case, the fact that it happened was mostly due to the efforts of a few people, who stood behind the idea and wanted it to happen. Pepsi-Cola Company had secured a pavilion in Osaka, Japan. Pepsi's marketing strategy being youth, it had planned a world-wide rock concert, with the theme "World Without Boundary." Somehow, David Thomas, then Vice-President of Pepsi-Cola International, seemed not to have been convinced that a rock concert was THE thing to do. Thomas went to visit his neighbor, Robert Breer, an artist who was making motorized sculptures. Breer, at first quite reluctant to become involved with a commercial company, eventually became excited and began to develop some ideas of his own. At this point, however, he contacted Billy Klüver, whom he had come to know from the *Nine Evenings*, and subsequently suggested to Thomas that Pepsi-Cola bring in E.A.T. Although Klüver was apparently very hesitant to become involved with a commercial venture, what convinced him to put forward a proposal was the idea that a group of artists would collaborate on developing an initial design and then be joined by other artists and engineers for the further development and execution of the project.[76]

Among the first artists consulted were Forrest Meyers, David Tudor and Robert Whitman, all of whom remained with it until the end. The development of the whole process of research and technical accomplishments, being a meeting of minds as well, can be followed in the book that was published. Especially Nilo Lindgren's essay is illuminating. Herein, we are informed that the technical complexities had been greatly underestimated, although they led to a number of interesting designs. A cloud surrounding the *Pavilion,* developed by Fujiko Nakaya, was generated by an atomizer system, "the largest pure water vapor system ever constructed." A complicated sound system was constructed for "automated control which could be overridden by live manipulation." A *Mirror Dome*, ninety feet in diameter, had "the unique optical property of reproducing an object or a person as a real image in suspended space." A *Clam Room and Laser Deflection System* had "a 'lightfoot' floor which illuminated footprints."[77] Outside the *Pavilion* were Robert Breer's "Seven white, dome-shaped battery-operated floats (that) moved about the plaza in front of the *Pavilion.*"

Although most of the participating artists came from the East Coast, the major technological contributions were developed in California, an aspect that has not been given much attention. The research for the construction of the *Mirror Dome* was done by Elsa Garmire, a physicist at the California Institute of Technology in Pasadena. She is currently a professor at the University of Southern California, Los Angeles. Garmire was connected with Ardison Phillips and David McDermott, two environmental artists who were actively involved with the art and technology group here. Under the management of Phillips, the Los Angeles group proceeded to develop a full-scale inflatable mirror structure, with a water-filled ring to keep it down, and finally built it in an hangar of the Marine Corps in Santa Ana, in the summer of 1969.[78] Fujiko Nakaya's *Fog Sculpture* was also developed in Pasadena by scientist Tom Mee, a cloud physicist. E.A.T.'s files show that the project was subject to many delays and technical failures, indeed, and largely overdrew its budget.[79]

From Klüver's introduction to the book it is apparent that he considered the whole development and production process a part of the project; in fact the most important part, for it was during this process that the exchange of ideas and experiences between the participating artists and those from other disciplines took place which constituted the educational basis for future experiences. This is quite important a realization in respect of the forthcoming changes in public art forms. The *Pavilion* itself was in effect a continuation of the *Nine Evenings,* on an enlarged scale and on a much more sophisticated level. As he wrote: "The organization and process of doing the project was in fact a 'mirror image' of the results we wanted to achieve, and was based on collaboration, personal responsibility, and experimenting. Most of the work on specific elements was done through collaborations which involved over 75 engineers and artists, and industries in Japan and different parts of the United States."[80] In fact, the project was technically not at all that advanced when compared to other pavilions. As an artistic experiment, though, it can be considered a success, and according to Klüver deserved to be treated as an art work.

In the case of the *Pavilion,* he therefore suggested to Pepsi-Cola to officially recognize the total work as an art work, in order to give it a legal structure. In a letter to Donald Kendall, President of Pepsi-Cola, Inc., he wrote "Our legal relationship to Pepsi Cola has developed so that the artists are put in the category of commercial artists designing a commercial product. One consequence of this is that we must obtain rights from all artists and engineers and others involved, particularly with regard to use of the *Pavilion* after Expo '70. Of course, there is no question of Pepsi's ownership and right to use and exhibit the *Pavilion.* Our dilemma is whether the artists have created a work of art or a work of commercial art to which there are rights which must be guaranteed... A decision to recognize the Pepsi Pavilion as a work of art and to treat it as such will set a much needed precedent in this area."[81] Pepsi-Cola never took this step and eventually the *Pavilion* was left in a state of gradual desolation and decay. This was certainly due to the fact that the relationship between E.A.T. and Pepsi-Cola had considerably cooled down, to the point that the company, the sole sponsor of the project, withdrew its support when E.A.T. presented a maintenance contract for $405,000, instead of the proposed sum of $185,000.[82]

Writer and art critic Calvin Tomkins felt that E.A.T.'s technology was particularly successful this time, but that Klüver's insistence on experimental programming in the *Mirror Dome*, and the ever-escalating costs of such experiments, had eventually exhausted Pepsi-Cola's tolerance for revolutionary processes. Jack Burnham suggested not long ago that E.A.T.'s gradual loss of its image as a corporate mediator might indeed have been precipitated by this unfortunate affair. He added that it was also because he felt that it had become an 'elitist' organization, by which he meant that it predominantly favored prominent New York artists by then.[83] True or not, a number of things caused the foundation's gradual disappearance from the center of attention. Klüver himself, although remaining active, took on a more modest role. Rauschenberg had already gone on to other things, as we saw. In addition, as said before, Klüver's interests shifted to promoting projects that lay outside the terrain that was traditionally defined as art; projects that were considered too social by the art scene. Artists were afraid they might become social workers.

Interdisciplinary Center for Advanced Visual Studies

While E.A.T.'s starting point was similarly utopian as that of The Center for Advanced Visual Studies, in that both organizations sought to participate in restructuring society through interdisciplinary public and communal art projects, one can distinguish a difference in emphasis: E.A.T. focused on the collaborative aspect between artist and engineer with a specific problem/project in mind, whereby both would have an equal share in the solution; the Center's emphasis was more on research, exploration, exchange of ideas and creativity rather than the finished art work. The Center for Advanced Visual Studies was founded in 1967 by György Kepes as the result of his efforts at MIT to connect the visual arts with the sciences.[84] A small 12,000 square foot building opposite the main building at MIT was erected, a rather unobtrusive building, barely visible among the other departments of this preeminent school, in which the Center has been operating until the present day. It took some effort to convince the Boards and find the financial support from other foundations, but Kepes persevered, and throughout the years, was able develop a certain status for this center, which is unique in its kind.

He described the origins of his ideas for the CAVS extensively in publications and interviews, from which we can deduct that the ideals and the objectives of the newly founded Center were connected with his vision perceiving a different future function of the artist in society: creating art forms that would further a better balanced and aesthetically inspired environment. But this would require a new understanding of the outside world by the artists, which could only grow if the artists themselves were to grasp the world of the scientist and the technologist better, which in turn would imply an active interaction and exchange of ideas. Kepes: "It seeks, above all, to develop new objectives. These are envisioned as complementaries of both the private and the civic sector of art: intensifying the infra-individual world and at the same time developing networks of communication between individuals, and between the individual and the environment."[85] Consequently, he defined his intentions and goals for the center: "The Center is planned first as a clearing house of artistic tasks that have authentic roots in our present condition. Second, the Center should be a laboratory where the most advanced technological tools can be tested for their applicability in the newly emerging scale of artistic tasks. Third, and above all, the Center is intended to become a testing ground for the needed collaboration among artists, scientists and engineers."

Apart from providing a learning experience for artists, he hoped that the scientists' contacts with artists would also lead to insights on their part not otherwise reached, for he truly believed that visual and aesthetic experiences could also serve scientific and technological discoveries.[86] Beside these general outlines there was no specific program or strategy developed to execute them, although Kepes articulated his design options in seven points in a 1968 article written for *Artscanada*, which encompassed the creation of a new environment which accounted for social and environmental concerns as well as aesthetic dimensions, seeking "a systematic approach to solve practical problems through the use of the new technological thinking and collaborations with sophisticated industries."[87] The Center for Advanced Visual Studies moved on various levels to realize its ideals. On a public level, there were the communal projects and exhibitions with which the Center presented itself to the outside world. The first public project that the fellows undertook under Kepes's supervision was the *Boston Harbor Project*. Intentionally interdisciplinary, it was set up to engage artists, structural, electronic, computer, and systems engineers, a city planner, a psychologist and a sociologist.[88]

An extensive description published in the same article reads: "The fellows of the MIT Center for Advanced Visual Studies, now engaged in preliminary exploration of a major project, recognize objectives similar to those of other collaborative groups: absorption of the new technology as an artistic medium; interaction of artists, scientists, engineers, and industry; the raising of the scale of work to the scale of the urban setting; media geared to all sensory modalities; incorporation of natural processes, such as cloud play, water flow, and the cyclical variations of light and weather; acceptance of the participation of "spectators" in such a way that art becomes a confluence rather than a dialogue."[89] It is a very general survey of purposes, giving only an indication of the type of work to be expected: multi-media light spectacles. György Kepes's personal proposal envisioned, for example, "a programmed, luminous structure floating in Boston Harbor, animating water and sky, (which) could welcome the visitors arriving by sea and air and at the same time give the inhabitants of the city a truly twentieth-century reminder that in spite of all our man-made wonders, nature, the sea and sky, are still with us."[90]

Kepes had invited seven artists to participate in the project: Jack Burnham, Gary Rieveschl, David Tulbert, Ted Kraynick, Otto Piene, Takis, and Wen-Ying Tsai. All proposed projects were to be placed on and in the water, and utilized kinetically moving objects, light and sound, either interacting with the movement of the sea or the city and harbor environments. The technological aspect generally involved computer programming, controlling movement or light sequences, etc. Interestingly, all artists entered proposals independently. Even if the artists may have shared similar concerns, the project never reached the state of a unified design, and was never realized.[91] Jonathan Benthall mentioned a few other civic proposals, such as plans for new sewerage systems, promenades, parks, fountains, the exposure of functional water-purification and hydraulic sculptures, experiments with ice configurations, inflatable tubes and so forth. They suffered the same fate.[92]

Douglas Davis has remembered that "in the early years there was an evident emphasis on two points: common projects, which included a series of proposals to form a large-scale luminous environmental structure in Boston Harbor; and the importance of establishing strong, continuous dialogues with the scientific community of Cambridge."[93] From these comments it might seem that nothing much was accomplished. However, the Center was actively involved in the organization of exhibitions, lectures and symposia. Then there was the internal level of operation. The fellows were invited to work with scientists individually, exchange ideas and set up a collaborative situation. Sculptor Wen-Ying Tsai, for example, worked with Harold Edgerton, electrical engineer and the father of stroboscopic photography, on the development of his cybernetic sculptures. Otto Piene also worked with Edgerton on strobe-timed light sculptures. Vassilakis Takis researched the potential use of tidal power for his kinetic art forms using magnetic forces with Ain Sonin, a professor in mechanical engineering. Stan VanderBeek worked with computer scientists on his early generation of computer graphics and slow-scan techniques. The filmmaker also had a long-lasting working relationship with computer programmer and artist Kenneth Knowlton from Bell Telephone Laboratories.

In general one can say that the individual contacts between the artists and the scientists, the personal relationship, was the most important factor in determining the success or failure of a project. And this has generally been true for all endeavors that ventured to create a collaborative network between the world of art and technology/science, whether it concerned Experiments in Art and Technology, Inc. or the *Art and Technology Program* in Los Angeles. If we try to characterize the direction of the early art works that came out of the Center, we notice that many projects of these years explore qualities of the medium light, such as electric energy, vibration, wave patterns, oscillation, and so forth. György Kepes himself had been concerned with the possibilities of light, literal and symbolic, since his friendship with Laszlo Moholy-Nagy. It was timely to experiment with this medium, as its immaterial qualities could refer to diverse notions of energy, time-space, movement. Other media explored here, and generally not available to artists, were laser, holography and computer technologies. At first, artists with a wide variety of concerns were principally attracted to the Center because it provided access to the latest technologies and offered a frame for communication and exchange of ideas between different disciplines. However, there also was the noticeable shift in interests since the early seventies, desrcibed above. Gradually, the use of advanced digital and electronic computer technologies - in video, music, holography, graphics, urban design, and so forth - were combined with research into natural phenomena such as solar energy, astrophysics, geology and biological systems, with the intention to re-establish a new relationship between mankind and nature. This shift can be seen in the kind of artists that were invited to the Center, as well as in the exhibitions organized between 1970 and 1975: *Explorations* (1970), *Multiple Interaction Team* (1972), and *ARTTRANSITION* (1975).[94] The exhibition included predominantly temporary works with an inclination to connect technological advancement with concerns about nature. The catalogue contains an introduction by György Kepes, statements by the artists, and short descriptions of all the works. Whereas in *Explorations* there still was a vested interest in the investigation of light media such as laser and holography, in cybernetic devices, and electronic and computer controlled systems, by 1972 the concerns of the artists had moved beyond the technological to create situations in which issues concerning the environment, ecology or nature at large were brought forward. A few statements by the artists themselves from the *ARTTRANSITION* catalogue give a more specific picture of the nature of some of the works. One should bear in mind that many of the works do not exist any more, so apart from photographs one has to depend on existing descriptions in the catalogue and reviews to get a picture of the nature of the works. The titles indicate that the majority of artists were exploring different kinds of light, the nature of heat, and air or movement, using some of the latest industrial materials and technological inventions.[95]

György Kepes designed "*Flame Orchard* (consists of) a number of 6 foot long, 4 inch diameter gas-containing aluminum tubes closed on both sides with 40 holes at the top where the gas jets are lit. A loudspeaker with an elastic membrane is mounted inside of each tube. Gas contained in the tubes is brought into vibration by the loudspeakers which transmit sound patterns composed for the piece and thus modulate the flames of the escaping gas." The piece was a collaborative project between artists György Kepes and Mauricio Bueno, composer Paul Earls and physicist William Walton. Otto Piene presented "(a) glass body consisting of panels with external large prisms and circa 440 multi-faceted small prisms inside; lights attached causing environmental spectral projections in a darkened space with white walls." Alan Sonfist defined his piece as "an ecological landscape which exhibits the natural conditions that existed on that spot before organization and planned landscaping of the city." Lowry Burgess wrote the following description: (excerpt) "The *Star-Pits* juxtapose two immense time-space structures, the celestial and the geological. I dig pits into the Earth to a layer at a depth equivalent in time to the age in light years of the light coming from two light sources: the globular star cluster in Hercules, some 22,000 light years from us; and the Great Spiral Galaxy in Andromeda, approximately 2,000,000 light years from us. ... The pits are entered by long ramps or grooves in the Earth. Down these grooves the Sun enters the pits marking the boundaries of night at sunset and dawn. The walls of the pit are carefully cut to reveal the layering of the Earth. The floor of the pit is mirror-water and placed upon it is a large water prism. The water in the open water prism blows breaking spectra upon the fresh earth walls. ... On the floor of the Star Pits is a series of lines and shapes which depict the proper motion of the stars. ... These lines show the shifting relationships between the stars in the period of time equivalent to the distance."[96]

With *ARTTRANSITION* the CAVS initially continued this direction under Otto Piene, who was appointed director in 1975. The focus was on thematic issues such as the current state of the interaction of art, science and technology, developmental media work, environmental concepts, communication and education, and "celebrations," Piene's special interest.[97] Lowry Burgess' and Alan Sonfist's works were indicative of a shift in the interests of a number of fellows at the Center, whereby interest in technology and science now was connected with an interest in natural systems, growth processes and natural materials, a direction that has grown stronger over the years. Even if light was still the predominant element in the installations, there was less emphasis on its technical features. Rather, attention was paid to including aspects which could give an environmental, ecological or social function to the art work. These ideas had in fact been part of György Kepes's goals when he founded the Center, but it was only in the early seventies that they bore fruit among a small group of fellows. The change was indicative of the new consciousness of the dangers, no longer potential but real, of unlimited industrial growth for the earth and mankind.[98] Otto Piene tried to continue the idea of public presentations, in the form of what he calls "Celebrations," or "Sky Events." Apart from the large collaborative and interdisciplinary *Centerbeam,* presented at *Documenta 5*, Kassel, 1977, these events have drawn little attention from the art world, except from insiders. For, as was the case with E.A.T., the CAVS received most attention during the late sixties and early seventies.[99]

A survey of the critiques of the CAVS idea(l)s and activities presents a mixed bag of positive and negative interpretations. While the initiative was received with overall acclaim, the projects were not, for the simple reason that they did not deliver what they professed. Jonathan Benthall analyzed the problematics involved in such projects as the *Boston Harbor Project* as follows: "Kepes, it is clear, is personally disappointed that his vision of urban art on a 'civic' scale has proved so hard to realize. The problem is twofold. First, there are the difficulties of finance and organization, especially during America's recent economic recession. Any such projects are bound to be costly and complicated, while most urban authorities' notion of 'art on a civic scale' is still a Henry Moore perched on a windswept piazza amid tall glass office-buildings. But, secondly, only a minority of good artists are interested in such projects. It is not merely that there has been a recent vogue for gallery installations about 'art about art,' but artists are understandably keen to maintain their independence from institutional bureaucracy."[100]

We have already mentioned the fact that a comprehensive unified program was never developed, which may be an indication of the difficulty in setting up a project in which artists not only collaborate with specialists in other fields but with each other. Emily Wasserman also commented that the artists were not or rarely willing to give up their artistic individuality and authenticity yet.[101] We have seen that Klüver had encountered the same problem during the organization of his *Nine Evenings.* In addition to the organizational and financial problems, one should probably add that the Center had little or no experience in dealing with public administration, legal authorities, civic organizations, and so forth. They were in no way equipped to even develop the interdisciplinary structure that is necessary for the development of large-scale communal projects such as the one proposed. Thus, the fund-raising to support the project probably collapsed not only because of the 'recession,' but also due to inexperience.

Jack Burnham, one of the fellows and participating artists at the time, voiced a different point of view in a more recent article, which rather refers to problems concerning the internal structure. He remembers that "during my first month and a half we met twice weekly to discuss Kepes's ambitions for erecting a colossol light tower in the middle of the Boston Harbor. Somehow the conversations and exchange of ideas remained maddeningly vague. I began to ask specific questions: Did the Center have funds for such a project or any idea of costs? No. Given that the Boston Harbor was directly in the flight patterns of Logan Airport, had the Center checked on the feasibility of the project with the local Civil Aeronautics Board, or with the Boston Harbor Authority? No. .." Burnham concluded that "due to the fact that the Center had been publicized, by virtue of its relation to MIT, as a technological nirvana for the artist, I found the situation mystifying. Slowly it began to dawn on me that the center's underlying purpose was not primarily to do visual research or to make art, but to produce lavishly illustrated catalogues and anthologies that impress foundations."[102] Burnham also related its "failure" to the fact that United States had meanwhile become involved in the Vietnam War, for which popular support was diminishing rapidly, and that, congruently with it, the uncritical admiration of technological progress dwindled away. That the heyday of technological progress was over and replaced by growing criticism, coupled with a new environmental and ecological concern.

Kepes's catalogue introduction to *Explorations* contained an emotional plea for a socially and ecologically responsible art. But Emily Wasserman remarked in her review, correctly it would now seem: "The show, like almost every art-and-technology exhibit mounted during the past few years, failed to end up looking as significant as its possibilities of cooperative design, vision and serious communication suggested. ... Too many of the pieces simply dwell on the exotic effects of the new media - magnetism, fluorescent substances, electronics, stroboscopic or laser light, etc."[103] Kepes's plea seems somewhat misplaced in comparison to the exhibited works now. If the civic projects that Kepes suggested were actually quite wonderful, they were so large in scale and so complicated that to accomplish even a small part of its undertaking would have been difficult for the Center of Advanced Visual Studies.

As a European, one wonders why these proposals and projects had to be so large in the first place. The early earthworks have the same quality. It probably sounds very simple, but I suppose that Americans do approach things on a bigger scale, they tend 'to think big.' The country has some of the world's grandest nature and there exists another concept of space. In any case, the role of the artist in all this was to be one of education and development of art forms with environmental dimensions. In defining a new social and political role for the artist, his vision was quite advanced, but very few artists were ready for such a change, i.e. ultimately to give up the authority of the art work to a large extent.

EAT/CAVS: Continuing Developments

Experiments in Art and Technology, Inc. still exists, with more or less active chapters operating in different regions. Attention from the art world has until recently been negligible. Apart from the reasons mentioned by Popper and Burnham for the fading away of the organization, which were basically criticism of the internal functioning of the organization, there may have been a series of external factors. The success of Experiments in Art and Technology encompassed the few years when the Art and Technology Movement was new, and quickly infiltrated the art world with a number of large exhibitions which in turn received a lot of attention from the (art) press. Whether the critiques were negative or positive is not all that relevant. The interest of many artists who experimented with technological gadgets faded soon, as they returned to previous or other means of expression. A review of the participants in *Some More Beginnings* shows many names that are now totally unknown. Few of the listed artists still work with technologically based media.

That the demise of the movement was caused by the arising critical attitude toward technology and the popularization of the ecology movement is only partially true. Although ecological awareness, combined with a growing concern about possible dangers of the new technologies related to the Vietnam War, affected certain artists to change their motivations as well as the kind of research, in retrospect it would seem that only few artists were truly involved in a concern for the environment.[104]

Also, art itself was shifting attention again. The inter-media performances where many different media were used and which included a lot of random elements and often insulted the public, no longer held the attention of the spectator nor the art world. After a while it all looked boring and amateurish; a 'professional approach' was once again required. And indeed, the performances of the later seventies became highly professional multi-media theatrical spectacles, in which nothing was left to chance. Renewed professionalization was required now. In addition, the digitizing of the electronic media such as video and the computer happened so rapidly that they required at least some technical knowledge. Artists had to educate themselves and specialize anew, if they wanted to communicate with engineers, video editors or computer scientists. What we see is a more specialized utilization of one medium instead of, say, ten. As computer technologies became more readily available from the early seventies onward, and computer arts groups were founded, E.A.T. became less involved with computer art, which had been one of its strongholds, thus losing part of its function. In addition, individual access to the industry had become easier, although only to a certain extent.[105] Frank Popper wrote in 1975 that "the E.A.T. organization, and their director Billy Klüver, seems to have finally shifted its emphasis to the engineer, to the detriment of artist and public. From matching artists and engineers, they have reached a point where they are encouraging them to go their separate ways, and the bulk of their publicity is now devoted to engineers. This may be the reason why artists have become estranged from the E.A.T. organization."[106] This tendency has only become stronger since 1975. However, as artists moved on to other directions, they did not stop using technology. It just became an integrated part of their work or activities. Nor did the collaborations between artists, technicians and scientists come to a halt. But instead of working under the umbrella of an organization such as E.A.T., the artists now preferred individual, personal relationships, whether they lasted for the period of one project or longer.

Concluding, one can say that Klüver's ideal of a network of artists and members of science and industry mutually exchanging ideas, with engineers developing projects together with the artists, and ultimately artists influencing certain technological or even scientific developments, just remained an ideal. This system of interrelations between the arts (mainly technologically inclined types of art) and the industries, not in a commercial sense but as an exchange of creativity, giving artists the power to directly partake in the industrial process and influence the directions to follow, "to make a difference," to quote Gregory Bateson, and which in Klüver's eyes meant not so much that art be given up, but that art should acquire some influence on society's structures, never happened. Although the efforts of Klüver, Kepes, and others, like Maurice Tuchman, never did come to fruition, from now on artists were able to handle their own affairs better, and search for the person or contact they thought to work with best. It may even be a crucial factor. The development reports of the *Pavilion* clarify that it progressed most rapidly when the artists started working on a one-to-one basis with the engineers. From the *Report on the Art and Technology Program* in Los Angeles we can deduce that only those proposals ended in a positive result, where the artist had been lucky enough to strike up a personal relationship with somebody in the company of his choice. And there were indeed a number of long-lasting relationships between artists and engineers, such as the one between computer programmer/artist Kenneth Knowlton and the late Stan VanderBeek, video artist Nam June Paik and electrical engineer Shuya Abe, video artist Woody Vasulka and computer scientist Jeffrey Schier, to mention a few.

There is a tendency among most young artists at the CAVS to at least acquire the basic knowledge to handle sophisticated equipment, and to be able to relate to the technician or engineer what they want to accomplish. Others go beyond the basics and even specialize. Myron Krueger has a background in computer sciences and liberal arts, for example. Todd Siler has completed studies in neurology, and perceptual psychology in addition to his visual arts degree. Bill Seaman, who has mainly become known for his work in video, has also worked in music composition and computers. Interests in science and technology certainly never died, it just took new directions, in pace with the rapid developments in digital technologies.

The Center for Advanced Visual Studies has gradually changed directions under its director Otto Piene (1974-1994) and still exists, albeit with difficulty. The idea of art as a mode of communication, as a mode to transfer information, that is messages, runs through Piene's philosophy about the Center. As the science of information has been an important part of the research done at MIT, not in the least because of its relationship with the communication sciences and communication technologies, the specialty of the Media Laboratory, this should not come as a surprise.

When Claude E. Shannon and Warren Weaver laid the foundation for information theory in the late 1940s, information was still defined as "The information contained in a message unit is defined in terms of the average number of digits required to decode it." Herein, information as organized data (not the interpretation of information!) is a measurable quantity that can be described mathematically, for example, the amount of 'information' that is transmitted as 'signal' via a telephone cable, telegraph or radiowave, which is energy/matter. Current ideas perceive information as a fundamental property of the universe, as a physical property existing independently of our ability to understand or decode it [107] Gregory Bateson (1904-1980) offered yet another definition in the late seventies: "Any difference that makes a difference." Concerned with disciplines as diverse as the behavioral sciences, biology, and philosophy, Bateson had developed a new interdisciplinary way of thinking about the nature of order and organization in living systems, by discovering patterns common to different disciplines, "the patterns that connect." Basic to his construct was the idea of the understanding of information, which realizes that communication is based on difference. Current research into neural networks done at the Massachusetts Institute of Technology, also called Connectionism, actualizes Bateson's thoughts. He was one of the first scientists to synthesize cybernetic ideas with anthropological data back in the fifties. Speculative as it may sound, I noticed a similar mode of thinking at the Media Laboratory, MIT's information and communication technologies' research laboratory, during a visit in 1993. To support this notion I refer to Steward Brand's comment in his book about the Media Laboratory, where he says: "The definition I kept hearing at the Media Lab was (Gregory) Bateson's highly subjective one."[108] The focus is not so much to develop socially relevant public art projects any longer, but on the fellows finding a visual language of their own, to communicate and educate; in which the exchange of information and collaboration with scientists and engineers is still a part, but a means to an end instead of an end. For Piene the artist is a natural educator, whose function it is to communicate a (i.e. his) message widely.

Publicly, the CAVS presents itself with so-called "Celebrations," as Piene calls them, actually organized to create or further CAVS as a community of artists. One could say that it has turned inward more. This attitude, which seems to have become stronger over the years, might in the long run prove to be a failure. Part of this may be due to another development on the campus of MIT, in which the CAVS failed to partake. Wounded pride, perhaps? In 1976, MIT's Council for the Arts presented a report that argued that the visual arts, including the performing arts, were in need of new facilities. Then the idea came to bring all the different art-related activities - of which there were quite a few apart from the exhibition program - together in one building, the "Arts and Media Technology Building." Many discussions about the context and morphology of the building and its relationship to the other buildings, i.e. its functions on campus, followed. When Nicolas Negroponte, former director of the Architecture Machine Group, became involved ca. 1982, the process seems to have precipitated and the architecture bureau of I.M. Pei & Partners was designated to design the building. Meanwhile Kathy Halbreich, director of the exhibition program, had proposed that a number of artists should become involved to collaborate with the architect on the design process. (The history of this is documented in a book, *Designing the Wiesner Building.*) At any rate, among those not consulted was the Center for Advanced Visual Studies. The CAVS withdrew from the project. Originally they were to be the major occupant of the new building. This happened parallel to the emergence of the Media Technology groups, whose interests were in high-tech electronic art forms, as the primary clients-occupants. Their interests were well represented by Nicolas Negroponte and Jerome Wiesner. The result was the Media Laboratory, which is perceived by many as a continuation of the ideas of the Architecture Machine Group. "If dialogue was the connecting idea for the Architecture Machine Group," concludes Steward Brand in his book, "the primary theme and meanings brought about by changes in context at the Media Lab is conversation, with computers and through computers."[109] The Media Laboratory has not become a visual arts facility, although there are many artists working there and some enter via the Center for Advanced Visual Studies. The Media Lab, as it is called, has become a serious competitor of the CAVS, not least because it is definitely more successful in collecting money from sponsorships. However, the Center remains interesting both as a facility that provides the interested artist direct access to the latest technological ideas, even if these are at the Media Lab, and as a facility that offers fantastic interdisciplinary possibilities. Fellows like Lowry Burgess, Alan Sonfist, Keiko Prince, Joan Brigham, among others, could expand their interests in the earth and nature to the sky, and the universe at large, using existing and new knowledge of planetary and space research, astrophysics, etc. During the seventies and eighties these artists have individually developed a 'public art' form based on an interdisciplinary approach, which focuses on environmental and communal concerns.

Chapter 3

Art as a Set of Relations

Critical Theories

As said earlier, by 1970 the kinetically defined works, which still prevailed in the early sixties, had gradually given way to different types of non-object processes, natural and technical, including evanescent phenomena, transient images, as well as the environmental space and context. The formal qualities, which had been a determinant factor in the (stylistic) analysis of the object-type art works, were now replaced by non-formal ones sometimes called dematerialized. I mentioned that the experimentation with natural and technological media came in part from the artists' revolt against the existing commercial art structure, and resulted in the use of non-precious natural, technical or transient materials. In fact, new materials were just as quickly co-opted as they were introduced.

The purposeful denial of form by the artists in the late sixties, and their search for a new art form beyond the old boundaries, also forced the critic/writer to re-think his or her approach. In the introduction I also mentioned that at first such terms as anti-form, excentric abstraction, or anti-illusionism were introduced to define these new directions, and that Lucy Lippard and John Chandler introduced the term dematerialization to indicate art's general evolution away from the art object between 1966 and 1972, as an overall circumscription of such ephemeral concepts as time or space, process, invisible systems, situations, experiences. Actually, the general term dematerialization was already in use in the sciences to denote continuing research into the invisible structure of nature and the universe. Multi-media or mixed-media art was the definition under which many of the works produced under the umbrella of organizations like E.A.T. and the CAVS were presented. However, with the few exceptions mentioned, such as Lawrence Alloway, most writers and curators in the United States treated the different aspects, including the collaborative aspect, separately.

If according to Billy Klüver a project like the *Pavilion* should be perceived "as an open-ended situation ... in the scientific sense," leaving no other option than to interpret the work as an ongoing experiment including its development, research and production processes, as well as the viewers' experiences, few art writers/critics dared to go beyond an aesthetic judgement of the final result. Unfortunately, few have been able to see the *Pavilion*, and, as was the case with the *Nine Evenings*, for those who did it was easier to write about the flops and failures than about the whole process that led up to the final result, the experience. An evaluation of the project in the larger context from which it was developed, produced and received was still hard to accept. When asked how she felt about the work as art, art critic Barbara Rose found it difficult to answer. "There were no previous criteria to apply to analyzing the Pavilion as a total work of art. Moreover, my only dialogue for weeks had been with scientists and artists, and my vision might be clouded from being too close to the project. There were no colleagues to compare notes with, no art-world feedback whatsoever," she wrote afterwards.[110] Rose described her view on twentieth century art-and-technology relationships from an art-historical perspective and concluded that the tangible results of the *Pavilion* had indeed been considerable, such as international communication and interchange, the introduction of executives, technicians and scientists to the values and goals of art, and that it "provided a constructive alternative to the museum-gallery situation in the form of a community experience." This idea of a true collaboration, "a specifically American experiment in democratic interchange," was, according to Rose, obviously a major, if not the most important, aspect of the whole project, even if proved a long and painstaking process. For during the process the artists had to learn to replace competition and had to submerge their ego in the service of a common goal, while the scientists and engineers had to adjust to a different way of thinking as well, learn to appreciate a different set of aesthetics, instead of only funtional values. Rose perceived the *Pavilion* as providing a new alternative approach to art, such as outlined by an artist like John Cage, who said repeatedly that art should interact with life again, and that the traditional boundaries between space and time arts should be dissolved. She also referred to the ideas of Marshall McLuhan and R. Buckminster Fuller. Yet her approach remains linear, in that she gives a chronological survey of relationships between art and technology. Although very informative, it does not attempt to suggest cross-disciplinary influences between current thoughts in the sciences, technology and the arts.

Writer Gene Youngblood, who was able to see the work personally, tried to analyze how technology affected the aesthetics of a work and the way it was perceived, introducing a language that was also used in systems analysis. Youngblood was not so interested in evaluating art-historical connotations. Rather, he concentrated on the new technologies used in multi-projection environments, and in experiments with image-making technologies like television, computer films, and so forth, and how they were about to create a new perceptual language in a changing context of what he called (after McLuhan's view) "the global intermedia network."[111] The majority of the reviews of the Los Angeles *Art and Technology Program* compared input (the enormous expenses and the collaborative efforts) with output (the art works), and arrived at the general conclusion that most of the works did not deviate much from what the artists had been doing all along. So why all this fuss about technology. In her evaluation of the program, Jane Livingston even expressed her doubts about the effects of longer-lasting collaborations in the future. She felt that "various experimental interchanges in Art and Technology were on the whole a polymorphous, discursive and non-organic process," and that these relationships were "intrinsically non-organic ones at least on an 'a priori' basis."[112] It may indeed have been true overall for this particular project, which was instigated by a museum and was therefore somewhat artificial in nature. The individual experiences, as well as the current developments, show a different picture.The whole development and production process, which was so important to the artists, was never mentioned in the reviews. Nor was the fact that works using advanced technologies such as laser (Rockne Krebs), mirror optic systems (Robert Whitman), gas plasmas (Newton Harrison), ganzfields and anechoic chambers (James Turrell/Robert Irwin/Ed Wortz) could never have been made without the assistance of engineers and other specialists. Newton Harrison and Robert Irwin even admitted that the *Art and Technology Program* contributed largely to the new direction their career took.

How difficult it was for writers and critics to assess these works beyond the context of art, is evident in a preview of the *Software* exhibition, curated by Jack Burnham for the Jewish Museum in New York (1970). In this exhibition he attempted to realize his utopian ideas about new interdisciplinary art forms in an environment governed by information and communication processing systems; technologies that in his view would change the knowledge system of our society.[113] Art critic Bitite Vinklers commented: "It seems that in many cases in Software the works are closer to being sophisticated technological facilities than media, or, to put it in different words, they imitate the services produced by the second machine age - polls, news services, television, and so on. That is, the artists are presently working at a level of electronic technology where they are using the computer, for instance, in exactly the same or almost the same ways as it is being used for practical purposes. ... Of course, the artists' purpose may well be precisely the kind of identity of practical and aesthetic functions just described, and consequently the complete integration of the work into one's environment - which in this the case is the information-processing aspect of one's environment - to the extent that the art work is no longer noticed or identified as such."[114] Vinklers wondered what might be so desirable about "this kind of art." *Software* turned out to be quite controversial, not only because, once again, the technology that was in fact the subject of the show failed, but also because one wondered why certain artists who had not shown the least interest in computer technology before were invited. Jack Burnham had invited a wide range of artists together with a number of scientists and engineers working in the field of communication technology, but few of the artists were familiar with the technological aspects, let alone with the wider issues at stake.[115] As a matter of fact, the majority of which were conceptual and idea artists. These artists proposed projects dealing with issues of language, time, space or the environment and the availability of a time-sharing computer network, and did not necessarily make the works cybernetic in the strict sense. The pieces were mainly restricted to the investigation of the body or the self, and stayed within the realm of Conceptual Art, while none ventured into any interdisciplinary forms of research. More related to the concept of the exhibition and indeed more interesting in terms of interdisciplinary systems research or communication issues at large, were the contributions of Hans Haacke, Nam June Paik, Les Levine, and the non-artists (groups) including Theodor Nelson and Ned Woodman (Labyrinth: An Interactive Catalogue), and The Architecture Machine Group from MIT, headed by Nicolas Negroponte. These installations were at least attempts to relate the existing concept of art to concerns that went beyond its boundaries, in the sense that they moved into the world of other (mass) media and disciplines, literally as a mode of research. Hans Haacke's contribution and that of Theodor Nelson & Ned Woodman, for example, created a participatory research system which dealt with the creation and distribution of factual information. Haacke had developed his *Visitors' Profile* with the assistance of Scott Bradner (Art & Technology, Inc., Boston) and Digital Equipment Corporation. A computer compiled the information and processed the received data for a statistical profile of the exhibition's visitors. In addition, he had wanted to present *News* (1969), a type of newsroom where local, national and international news was received via teletype print-out just as it was received at the real news services.[116]

Labyrinth was an interactive program which presented the viewer with personalized information as well as a print-out with the information requested about the exhibition related to his or her special interests. One could then add this information to the catalogue, making each catalogue a different, individual copy. The computer also had a conversational program, so that visitors could retrieve information, ideas and opinions of previous visitors and offer a response, which then could again be used by the next user.[117] The most controversial project dealt with the possibilities of artificial intelligence and learning machines. The Architecture Machine Group had developed *SEEK*, "a mechanism that senses the physical environment, affects that environment, and in turn attempts to handle local unexpected events within the environment." In a plexiglass box, gerbils were 'playing with' small aluminum blocks which were subsequently carefully re-arranged into symmetrical structures by a programmed mechanical arm. The idea was that the robot/architect/arm should 'read' or learn to predict the unpredictable behavior of the gerbils.[118]

Jonathan Benthall, who had been publishing a feature about the art and technology phenomenon for *Studio International*, considered the project valid as art but insisted on making a distinction between artistic and scientific experimentation. As Benthall later on evaluated: "The computer and allied devices will certainly continue to be used in art. Works like SEEK point forward in new directions, where the computer is treated as a medium and its social role is explored."[119] How he envisioned this role and what its consequences might be, he did not detail. Perhaps, it was just too early to tell. From the artists' statements we can deduce that only those artists who were interested not only in utilizing the computer, but also in issues concerning information, communication and media at large, were the ones who were familiar with the ideas and the language of cybernetics and systems theory.[120]

One might say that on the one hand, it seems that the artists invited, with a few exceptions, were not ready to perceive the artistic contribution they could make to these new technologies by posing the right questions about control, future utilization and development, and so forth, and that therefore their work could not go beyond an investigation of the technical aspects. After all, one has to familiarize oneself first with any technology, understand the ramifications, before one can extend its commonplace functional use into an artistic one. On the other hand, it is not just likely but certain that these works, representing different values altogether, require the new aesthetic vision which Jack Burnham and Gene Youngblood advocated.

On the whole, it is striking that among the early writers venturing to write about the utilization of the computer in art and its implications as communication and information medium there were so few art historians and art writers. Nearly all the contributions came from scientists or professionals in the field, and occasionally, artists working with these new technologies. The lack of information provided by the existing art magazines and the relative inaccessibility of scientific and technical periodicals for non-professionals such as artists were no doubt important considerations in the development of the concept for *Leonardo* in 1968 by artist-scientist Frank J. Malina[121].

New Concepts in Science

In respect of the interest in scientific and technological contexts by artists, it is illuminating to insert a brief survey of the fantastic changes in the sciences that occurred earlier in the century, highlighting those ideas that appear to have interested artists most. The discoveries in the physical sciences since the nineteenth century forced physicists and other scientists to review the mechanistic concepts - also called Cartesian paradigm - that dominated our vision of the universe. This deterministic view of the universe defined real(ity) as that which can be analyzed or explained scientifically by a set of procedures combining empirical experiment and quantification. This vision was based on Isaac Newton's universal laws - i.e. his law of gravitation, and law of motion - published in 1689 in the *Principia Mathematica.* Herein he mathematically described the fundamental, which means in Newton's context absolute, relationships between physical phenomena, such as the orbits of the planets around the sun and of the moon around the earth, or how all bodies fall at the same rate. The most influential philosophical understanding deduced from these laws was that the universe functioned as a clockwork, a machine. Until Albert Einstein's general theory of relativity (1915) all scientific and empirical experiments were based on these principles. The general theory of relativity led to a completely new way of viewing time and space. In Newton's theory, time and space had been two absolute and unchangeable concepts. Einstein's discoveries connected time and space. In addition he showed that the time-space continuum was a relative dynamic constellation that changed according to ways in which either mass (bodies) or energy (forces) were affecting the environment (the curvature of space and time).[122] These discoveries have been excellently discussed in books on the physical sciences. Suffice it to say that Einstein never fully accepted the implications of his own scientific conclusions, since they, in theory at least, entailed the necessity to review the classical laws with which physical science had come to look at nature. Subsequent research in the world of sub-atomic particles brought in a whole new set of dimensions concerning reality, and it is not an exaggeration to say that the discovery of the quantum paved the way for one of the most important developments in the physical sciences of this century: quantum theory.

Its seeds were planted by Max Planck's discovery of the quanta about 1900. Quanta are a measurement for the energy or light that is radiated by electrons in packets instead of continuously, and each quantum has a certain amount of energy that increases proportionally to the frequency of the light waves. A major implication of this discovery was the indeterminacy of certain relations in the quantum world, which was postulated by Werner Heisenberg in his Uncertainty Principle (1927). This principle said that, when one tries to measure an elementary particle, one cannot know both its position and velocity at the same time; the more accurately one knows the one, the less accurately one can know the other. Werner Heisenberg belonged to a group of physicists, together with Max Born, Niels Bohr, Wolfgang Pauli, Pascual Jordan, among others, who worked toward a definitive and unified atomic theory in the 1920s. Two years earlier Heisenberg had formulated his matrix mechanics which explains different atomic light phenomena as parts of the same system, leading up to this discovery which would turn out no less than a change in paradigm.

Physicist Stephen Hawking has explained this succinctly: "In general, quantum mechanics does not predict a single definite result for an observation. Instead, it predicts a number of different possible outcomes and tells us how likely each of these is. That is to say, if one made the same measurement on a large number of similar systems, each of which started off in the same way, one would find that the result of the measurements would be A in a certain number of cases, B in a different number, and so on. ... Quantum mechanics therefore introduces an unavoidable element of unpredictability or randomness into science."[123] This also meant that the classical - Newtonian - notions of causality and determinism had become obsolete. In addition, Heisenberg's Uncertainty Principle showed that the observer/researcher always influenced the outcome of the experiment, making the idea of an independent, objective result obsolete; that is to say: in the quantum world. The lay-reader should be aware of the fact that quantum physics was in essence a world of mathematical formulae, a mathematical construct, for which new visualizations and descriptions had to be found. For the quantum world is a micro-world, a world invisible to the human eye, where different rules govern.[124] However, the influence of quantum mechanics on other scientific disciplines, such as chemistry and molecular biology for example, as well as on many new technologies, has been immense. We should keep in mind that its origins lie in Maxwell's theory of electromagnetism, which posed a mathematical equation from which resulted that electricity and magnetism were two manifestations of the same force, and which proved that light was nothing but a special form of electromagnetic radiation, subsequently described by its complementary functions of wave and particle. It is generally accepted that the developments of the electronic and digital technologies in the twentieth century are in essence a transformation of this most important nineteenth-century discovery. Its importance is based on the fact that its mathematical equations were found technologically applicable.

Meanwhile, another direction in physics had continued research into the concept of energy. It is known that all changes in a certain situation, or physical condition - over a period of time - require energy, that is an exchange or loss of warmth takes place. The first law of thermodynamics had shown that no energy is ever lost, just transformed, moving again toward equilibrium. However, the second law of thermodynamics asserted that the entropy of an isolated system increases with time, or remains constant, for a reversible system. Ludwig Boltzmann introduced the concept of entropy as a statistical formula into the science of thermodynamics, stating that natural processes, according to the second law of thermodynamics, always move into a direction of growing chaos, whereby entropy is a measure of randomness, of disorder. The concept of entropy has remained a much discussed topic throughout, particularly in connection with the rising ecological concerns and speculations about a possible heat death of the earth. It was very attractive to artists because of its connection with ideas of randomness and disorder.[125]

Yet another concept arose from the discoveries made in the twenties which was to become an influential new model of description, interpretation and analysis of at first biological phenomena, and which was consequently extended to bridge the growing specializations and function in an interdisciplinary manner: systems analysis. Ludwig von Bertalanffy, whose main fields of interest were biophysics and related work in cellular and comparative physiology, had come to the conclusion that the narrow boundaries of specializations and subspecializations hampered continuing research and the exchange of information crucial to understanding cross-disciplinary research. He thought the research conditions in the biological sciences very unsatisfying, for its use of elementary physical laws to explain vital phenomena proved to be totally inadequate for his purposes. His studies into living biological organisms had led him to the conclusion that certain questions regarding such living organisms could not be answered by the Newtonian cause-and-effect theories, which divided phenomena into elementary units, and that an explanation of the parts was not sufficient to explain the functioning of the whole. For, by looking at a phenomenon in isolation, one overlooked the fact that there always existed an exchange of matter, of energy either between the elements within the organism or between elements in the environment. This caused Bertalanffy to perceive the organization of living organisms, or animate matter, as an "open system", as opposed to a "closed system." Already in the late twenties he wrote: "Since the fundamental character of the living thing is its organization, the customary investigation of the single parts and processes cannot provide a complete explanation of the vital phenomena. This investigation gives us no information about the coordination of parts and processes. Thus the chief task of biology must be to discover the laws of biological systems (at all levels of organization). We believe that the attempts to find a foundation for theoretical biology point at a fundamental change in the world picture. This view, considered as a method of investigation, we shall call 'organismic biology' and as an attempt at an explanation, 'the system theory of the organism.'"[126] He introduced the concept of the system as a means to describe the different related energies at work. Meanwhile Ludwig von Bertalanffy had come to yet another insight: that similar general conceptions and viewpoints were being developed in other disciplines, but that a language that researchers could understand and work with as a medium to exchange this knowledge and information was lacking. He came to the conclusion that a new theory of scientific concepts needed to be conceptualized, which he called General System Theory. Its subject matter was (and is) the formulation of principles that are valid for systems, whereby systems were defined as "sets of elements standing in interaction," in general, whatever the nature of their component elements and the relations or "forces" between them. The foundation for the General System Theory was first formulated in 1937 at a seminar conducted at the University of Chicago, but only published after the Second World War.[127] Inter- and cross-disciplinary research became a basic feature of the new science, which was based on the premise of 'seeing things in relations.'

The discoveries in quantum theory have only supported these ideas, and the recognition that the analysis of natural phenomena as relatively autonomous parts wanted replacement by an analysis of their functions, but in interaction. The primary emphasis was now on "undivided wholeness, in which the observing instrument is not separated from what is observed."[128] Max Planck also sustained this point of view when he wrote, in 1933: "It is impossible to obtain an adequate version of the laws for which we are looking, unless the physical system is regarded 'as a whole.' According to modern mechanics (field theory), each individual particle of the system, in a certain sense, at any one time, exists simultaneously in every part of the space occupied by the system. This simultaneous existence applies not merely to the field of force with which it is surrounded, but also to its mass and charge."[129]

Jack Burnham: The Systems Approach

Jack Burnham played a prominent role in the art scene at the end of the sixties. Burnham started out as an artist but established his reputation, above all, as a writer/theoretician and organizer of exhibitions. From 1968-1969 he was a fellow at the Center for Advanced Visual Studies. In an article, "The Aesthetics of Intelligent Systems," he described how his residency allowed him to use the time-sharing computer system at MIT's Lincoln Laboratories. His fellowship at the Center obviously also allowed him to touch on the different areas of scientific and technological research whenever possible, and to participate in the exchange of ideas with scientists. His interest in the possibilities of information systems, computer networks or artificial intelligence was certainly stimulated by the research done at MIT. Here Burnham met, among others, Marvin Minsky, one of the pioneers in research on artificial intelligence, whose chess programs and ball-catching robots had received much publicity. He also became acquainted with Nicolas Negroponte and his Architecture Machine Group, and invited them to participate in *Software* with the installation *SEEK*, mentioned above.[130]

Writer/artist Jack Burnham belonged to those few who were in the vanguard to connect the twentieth-century developments in technology and science with the visual arts, in particular sculpture. He felt that the western world was changing from an object-oriented to a systems-oriented society.[131] Sculpture's change from objet d'art to système d'art paralleled, as he defined it, the intellectual framework produced by the scientific community. The traditional tools of art-history and criticism - whether it is a stylistic and formal analysis, iconographic, or the linear development model - could not explain these changes sufficiently and in *Beyond Modern Sculpture* and his subsequent writing he set out to develop an approach which presupposed twentieth-century sculpture as a system, in which changes showed a continuous correlation with those in science and technology. He met with severe criticism of his methodology from art-historical sides, which was to be expected. Burnham's attempt showed weaknesses, indeed, when it came to explain not so much the relationships but rather the dissimilarities that occurred between art and science during the first part of the century.[132] Most controversial, however, was his speculative prediction that the art object, being an inert artifact, would eventually disappear as a cultural expression to be replaced by what he called a "systems consciousness."[133]

His prophesy at the end of the book clarifies what he meant by that: "The stabilized dynamic system will become not only a symbol of life but literally life in the artist's hands and the dominant medium of further aesthetic ventures. ... As the Cybernetic Art of this generation grows more intelligent and sensitive, the Greek obsession with 'living' sculpture will take on an undreamed reality. ... The physical beauty which separates the sculptor from the results of his endeavors may well disappear altogether."[134] It is significant that he could only have developed these theories right at that time, allowing him to synchronize the new directions in sculpture with the ideas appertaining to the general system theory and its practical applications in computer and communication networks. Jack Burnham's ideas of a systems art were in fact derived from Ludwig von Bertalanffy's general system theory, which he referred to as well. During the sixties and early seventies the number of publications on systems analysis quickly mushroomed, and Bertalanffy's approach found a modified but wide application from the seventies onwards. Its basic ideas were generally accepted by different disciplines, one of which was the visual arts.

To summarize the motives for the development of the general system theory: it reciprocated the growing demand for the introduction of a generalization of scientific concepts, due to the fact that similar concepts and models had appeared in different fields, making a conceptual model necessary which would make interdisciplinary understanding and collaboration possible. It facilitated the introduction of new categories in scientific thought and research, with concepts like organization, directiveness, teleology, using mathematics and statistics as descriptive methods. Most issues nowadays require a systems approach, being problems of interrelations with many variables. One of the most important features of the introduction of these conceptual models based on mathematical formulae was that they were interdisciplinary, and transcended the conventional departmentalization of science. By now, a vast realm of systems technologies, related to technological developments, has arisen, ranging from information theory, control technology, graph and network theory, to automation, computerization, software programs applying systems, etc.[135] In Burnham's interpretation, the art object as the subject of analysis itself was replaced by the notion of a system, whereby a system was defined as "a set of relationships in action" (Ludwig von Bertalanffy). The elements of these relationships implied change(s) in time, that is process(es), which were interdependent and non-deterministic, however, since influences from the outside were not predetermined and could only be described scientifically in terms of possibilities.

Formal interpretation is based on the visible elements of a composition. In a systems context, equal attention is paid to invisible elements, to processes and changes in the work, as well as its relationship with the outside. These thoughts were the beginnings of what Burnham was to call a "systems esthetics." He proposed this term in opposition to Michael Fried's adjectives theatrical, literalist, and post-formalist esthetics, in his well-known article *Art and Objecthood*, in which Fried had comprised the materialist tendencies in modern sculpture into the one term "objecthood."[136] Fried thought that contemporary sculpture, being more and more a composition of different kinds of materials, was moving toward theatricality, a tendency he considered dangerous for the independence of art. For Burnham, the term 'systems esthetics' seemed to encompass the situation more fully, as it included the concept of the boundaries of art. For, when an art work crossed its boundaries, it became part of a context, or a system. Consequently, a systems esthetics was no longer solely limited to the specific concerns of the art world.[137] In Burnham's view, this meant that an analysis of these art works only made sense if one took into account "their assigned context." As context, he not only defined the immediate literal art space, but also the larger space of social, political, and technological contacts and collaborations with which the artist had to deal.

It was therefore not possible to regard a fragment of a work of art and attach higher or symbolic meaning to it, all system components being interdependent. Among the artists who had acknowledged this approach according to Burnham's thesis were, among others, Hans Haacke, Les Levine, Dennis Oppenheim and Robert Smithson. The following sections are revealing in that Burnham used the array of systems terminology to describe and interpret the work. Burnham: "If Smithson's *Site-Selections* are didactic exercises, they show a desperate need for environmental sensibility on a larger than room scale...Smithson has transcended this by putting engineering works into their natural settings and treating the whole as a time-bound web of man-nature interactions." Dennis Oppenheim's *September Wheat Project* was characterized as "Oppenheim is using the untapped energy and information networks of the day-to-day environment." In this and other sowing and harvest projects, searching the land, sowing the wheat or oat, "cutting, gathering, baling and trucking" were all part of the art process or art work, according to the artist and Burnham. *Levine's Restaurant* (1969), at 19th Street and Park Avenue, was described as "a self-organizing, data-gathering system," a "real time" process/ system.[138] In terms of art, the move from object to system implied for Burnham that the new art forms were no longer defined by a fixed shape or boundary, but behaved like a system that alters in time and space according to both internal mechanisms and external conditions.[139] He arrived at the conclusion that sculptors were not simply adopting new materials and new standards of fabrication. There was also a new aesthetic evolving from the synchronization with the technological and scientific ideas, for, as he explained, "the artistic concerns with man-made systems, processes, ecological relationships, and philosophical-linguistic involvement of Conceptual Art ... deal with underlying structures of communication or energy exchange," whereby computer technology would be "instrumental in redefining the entire area of esthetic awareness."[140]

As mentioned previously, in 1970 Burnham organized an exhibition with the title *Software,* in which he attempted to realize his utopian ideas about new interdisciplinary art forms in an environment governed by information and communication processing systems: technologies that would change the knowledge system of our society. The computer technologies would require new organizational structures for the exchange of information. These would no longer be based solely upon technology, but depend upon the knowledge necessary to have access to these tools; for those with knowledge and access would have the power. If artists wanted to participate in this environment, let alone exert some influence, they would have to learn about these technologies. Whether the result was called art or not was irrelevant at this point, according to Burnham. In fact, he thought that this realm would contribute to blurring the classical distinctions between art and non-art, between art and technology, or these distinctions might even become obsolete. Wrote Burnham: "As the computer environment further condenses the known world and as it increasingly becomes an elegant surrogate for global experience, a profound change will take place in the acquisition of knowledge and sensitivity."[141] His involvement with cybernetic ideas and self-organizing intelligent systems and the current optimistic belief of the researchers in artificial intelligence that it was possible to create systems with an artificial intelligence that could do the same things as human beings, that is have a life-like cognition, certainly influenced him to envision a situation where art and technology were advancing life, albeit from opposite directions, and that their meeting point would result in a resolution of art and technology: in the creation of life itself.

In a later article Jack Burnham set out to rectify this position, having come to the conclusion that it had "erred gravely in its tendency to anthropomorphize the goals of technology. ... What we are seeing when we view art is a fusion of cognition and gesture; as the historical semiotic of art evolves, this becomes increasingly apparent."[142] Of course, neither happened. Burnham's utopian idea about the future of art extrapolated too far from the existing situation. However, his connection of the changes in the visual arts with systems ideas still seems relevant, if restricted to those artists who sought a new context for art, and at that time. For it is certainly true that those artists who were involved with processes, events, time and space, and technical applications that involved interdisciplinary research, had a general knowledge of systems analysis. The use of cybernetic ideas, requiring scientific and technological knowledge, was found more among artists involved with computer-based art forms or sculptures, video, television, satellite and other communication media. The statements and writings of artists like Robert Smithson, Nam June Paik and Paul Ryan indicate that they were not only informed about these theories but could also incorporate the implications, artistically if not scientifically, into their art work. Robert Smithson systematically referred to the new thoughts on entropy in science and technology, and the application in his work.[143] In his contributions to *Radical Software* video artist Paul Ryan, in particular, described the utilization of the new artistic medium video in relation to the cybernetic and systems theories. He was active in the guerrilla TV scene for a while, then disappeared more or less from the art circle to purposively concern himself with social and, in particular, ecological issues. His writing is an example of an artist who was very knowledgeable about the cybernetic theories of Norbert Wiener, the media theories of Marshall McLuhan and so forth, interpreting the portable video system as an instrument to get access to information, to controlling the flow of information by means of "guerrilla warfare" systems.[144] What is significant here is not just the fact that Ryan was quite informed, but that he theoretically connected the development and function of a new medium, a new technology in the visual arts - video in his case - with these theories to create a theoretical construct, using guerrilla warfare as a model.[145]

Nam June Paik has also referred to the science of cybernetics, and in particular the new communication and information theories in relation to his own work in video, television and satellites in his writing. In a 1967 article, "Norbert Wiener and Marshall McLuhan," Paik set out to compare Norbert Wiener's theory of cybernetics with Marshall McLuhan's media theory.[146] A few artists perceived their work as cybernetic, such as sculptors James Seawright and Wen-Ying Tsai.

During the sixties systems theory and cybernetics rapidly spread among other disciplines, and its jargon was popularized, and also found its way into the visual arts. Many of Burnham's ideas (and use of language) about the consequences of the continuing infiltration of computer technologies into our daily lives can also be found in contemporary publications with a concern for the social, political, and foremost, human consequences of the computer technologies. Compare economist Peter Drucker's analysis, much read at the time: "The systematic and purposeful acquisition of information and its systematic application, rather than 'science' or 'technology', are emerging as the new foundation for work, productivity and effort throughout the world," when he discussed some economic and political consequences of the new computer technologies.[147] Or the economic theories of Alvin Toffler's *Future Shock*, and the ecological concerns of Charles Reich in *The Greening of America*, both published in 1970, as well as the media theories of Marshall McLuhan, already mentioned.

Marshall McLuhan's notion of the computer as an extension of the senses is met frequently, as well as the idea that we have become a society based on knowledge rather than experience. This led to the introduction of a whole terminology that came from these disciplines. It had given words like structure, pattern, organization, indeterminacy, interrelatedness, a new and specific meaning. Cybernetics introduced yet another new lexicon which was reflected by words like feedback, information, parameter, software, hardware, entropy. As said, an important aspect of both sciences was their interdisciplinary approach, and one can see that exactly this aspect would attract the artists who wanted to reach beyond art's known boundaries. That an infiltration of the ideas of cybernetics and systems thinking was facilitated by the experiments with new communication media, which were accompanied by artists' collaborations with scientists and engineers, should be seen as a logical accompaniment.

As I have said, artists were thus brought into direct contact with new technologies and scientific developments and theories. Thus, it was not only the systems terminology that was used as a way to describe the new directions in art. A group of artists truly related their work to the developments that were taking place in the sciences, even if the extent of their understanding remains difficult to measure. A main problem was constituted by the fact that the systems analysts and cyberneticians themselves applied different definitions. The terminology itself had varied meanings too, and was often used wrongly. Some artists emphasized the continuing abstraction of the sciences, as denoted by the growing importance of mathematical concepts or formulae as a conceptual framework and that of twentieth-century art as expressions of a similar development. Others perceived the new concepts of time-space, the changed subject-object relationship, indeterminacy and uncertainty as similar to those in the visual arts, although there was in my view no specific correlation between the developments in the visual arts and the sciences. Although a systems analysis in the narrow scientific sense proved difficult - Hans Haacke was the only artist who adhered to this definition for his art for a longer period of time - the multiple internal and external relationships of Haacke and other artists in search of another context and territory for art do require an interpretation that accounts for this 'organized complexity,' in particular since these works came to involve the knowledge and expertise of other disciplines, creating a complex layering in the content of the work of art.

The systems sciences and cybernetics have become accepted as valid methods in many disciplines, including the social sciences. Why this approach has been neglected by the scholarly art world is a question I feel incapable to answer. One possible answer is that the direction of the official art world moved toward the object again, while the artists who worked on and with the land moved in the opposite direction, showing little inclination to participate actively in exhibitions or other art activities. Another possibility is the fact that technologically-based art forms, with the exception of video art (which also remained marginal) disappeared from the main stream of attention around 1974, and computer art forms were systematically neglected by the art world. Also, other philosophical approaches, such as structuralism, followed by the post-modernist French philosophy advocated by Jean Baudrillard, Jacques Lacan, or Jean-François Lyotard, became en vogue. Interestingly, artists currently working in the field of interactive computer networks show a renewed interest in the systems theories and cybernetics of the late sixties and early seventies, now related to yet another wave of concern about the dynamics of ecological systems. This probably also explains current interest in the philosophy of Gregory Bateson.

The theories about a civic art form as proposed by György Kepes never materialized in the ways envisioned. Neither did the ideas of Billy Klüver, nor Jack Burnham's notion of the disappearance of the art object to assimilate with an overall 'systems' orientation as the dominant view of the world's organization, whether man-made or natural in origin. Yet one cannot do away with these ideas simply as idealistic concepts guided by an overall but vague call for social and political change of the times. Although the experimental, non-object, non-precious art forms, which were perceived as necessary indeed to break with the traditional concepts in the first place, still fitted the avant-garde idea of art as the expression of revolt and protest, a genuine belief that the traditional concept of art no longer had a function in a changing society - and the necessity for new forms made them search for possibilities outside the realm of the art system into what one might define as the environment at large.

Radical Software

"Electronic media have become looped-in to our neural networks. We need a minimum of information flow not only for physical survival, but also for psychological balance, because electronic media are as omnipresent as light" stated Michael Shamberg in 1971.[148]

Radical Software was published from 1970 to 1974 by Raindance Corporation, and founded by Michael Shamberg. The first edition was printed in July 1970 and reprinted as *Radical Software/Number One* in September 1970. *Radical Software* began as a "service to alternate television people," but with the advent of video technologies the magazine immediately followed "the trend towards alternate uses of technology and media. ... In fact, we think the combination of hardware and software trends (e.g. in addition to high access television, the availability of mini-computers will accelerate rapidly over the next few years) will see a whole generation of Americans coming to terms with its technology as a better tool for social change than worn-out, non-Cybernetic political models."[149] *Radical Software*'s prophesies did not live to see the technological models it promoted survive. Yet what was so significant about this magazine, was its interdisciplinary approach. Thus, the magazine was able to collect contributions by the foremost scientists and theoreticians in these fields, such as Buckminster Fuller, Gregory Bateson, Gene Youngblood, Warren Brodey, among others. Although the magazine was not nearly as much involved with the arts as with the social, political and cultural applications of systems theories or cybernetics within the realm of mass media and in particular television, there were quite a few contributions by artists, i.e. those artists who were interested in media and communication issues. The majority of them had recently started to use the new portable video technology. Around Michael Shamberg were Paul Ryan, Frank Gillette, Ira Schneider, and Andy Mann, all well-known video artists of the first generation. Yud Yalkut, Takis, Nam June Paik, Juan Downey, and Douglas Davis also wrote for it, to mention a few. The magazine was indeed well read among this group of artists, and quickly reached a circulation of 5,000. *Radical Software* was the only magazine which made the ideas of cybernetics, systems theory and their background known, and combined them with new theories about mass media and communication technologies, accessible and popular among artists. For publisher Michael Shamberg this approach was crucial because it was a logical consequence that the interdisciplinary developments in science would also be followed by a cultural shift: "The interdisciplinary approach in scientific and social problem solving is such an attempt at remodelling information access."[150]

Cybernetic and systems notions with their accompanying vocabulary were mainly applied to the possibilities of the new media systems, such as video, cable, satellite, etc. Words like system, feedback, information, software, parameter, entropy and negentropy, process, pattern became the principal vocabulary in the writers' argumentations. Shamberg pointedly noted these changes in linguistic terminology already in 1971. "The new lexicons reflect and generate cultural change. The 1960s, for example, were given their character partly by the widespread use of psychoanalytic terminology to describe practically anything. Already new terminologies are compelling for the direction of the 70s. ... Moreover, because videotape is part of a whole way of viewing and using technology, and because technology has until now been the province of scientists and government, by mixing technological jargon and slang we can make technological design responsible to a whole alternate cultural view. ... Unfortunately, most people from the alternate culture don't understand the dynamics of information."[151]

The subject matter dealt with was very diverse, traversing a wide range of disiciplines: biology, physics, anthropology, education, communications, and art. The topics were generally discussed in relationship to the new media and communications technologies. Although in particular the new satellite technology was a cause for great concern at the time, since it immediately brought to the surface the deeply rooted fears of an Orwellian world, dominated by surveillance stations and a centralized control of power, the reader may notice a predominantly positive, yet critical, attitude throughout the articles in the magazine. This may wonder, because the major change of opinion about the a priori ideas of progress in new technologies took place shortly after the magazine started as an alternative media magazine. The euphoric optimism of the early sixties with regard to the scientific and technological advances and in particular the computer-based technologies had gradually shifted and by the early seventies was critically connected with the 'automated battlefield,' as Vietnam was called, and its damaging effects on the environment.

Only one or two years into the seventies, the majority of the articles discuss man's relationship towards nature and technology or the environment at large, the cosmos. Many ideas expressed in these contributions evolved from the sciences, but are now combined with a critical and sometimes anti-technological attitude, and always show an ecological consciousness. The mood is obviously a completely different from the one encountered at the early Experiments in Art and Technology ventures, or the art and technology exhibition in Los Angeles. *Radical Software* witnessed the growing disenchantment with the official media, such as television, but still held on to a positive search for alternatives. Overall, there is a quest for a new balance, due to the growing concerns about the precarious condition of the earth brought about by modern technological progress; a quest for a new mode of thinking. In this, *Radical Software* is a forerunner of the nineties. I want to point out that the activists of the seventies who produced such magazines like *Radical Software* and the *Whole Earth Review* played a prominant role in the development of the current computer and communications networks.[152]

Leonardo

The magazine was founded in 1968 by Frank J. Malina, who had found a publishing partner, Robert Maxwell from Pergamon Press Ltd., Great Britain. In the first - January 1968 - issue Malina stated the aims and scope as follows: "Leonardo is intended to be primarily a channel for communication between artists." Each issue would have invited articles reviewing new developments in the *fine arts*, and new materials or technological and scientific techniques that might be of interest to artists. Malina also thought it important to discuss a wide range of topics from "the fields of physics, psychology, cinema, theatre, aesthetics, philosophy, architecture, etc."[153] There would also be space for letters, notes, book reviews, announcements of events, institutions and so on. Although initially all new developments in contemporary art were included, there was already a slight tendency to prefer new technlogical developments, in particular kinetic art forms, and interdisciplinary relationships between the arts and sciences or technology. It was this tendency that would become the magazine's main focus. Malina also stressed the international scope of *Leonardo* as the subtitle shows: International Journal of Contemporary Art. The first issues were both in French and English, and the main editorial offices were in France until 1983, when they were moved to San Francisco, California.

The directions which *Leonardo* took were obviously inspired by its founder, himself a pioneer in light and motion art. His contributions to kinetic art were recognized in the field, as reflecting the "new landscape" (György Kepes) provided by science and technology. Frank Malina's interest in interdisciplinary relationships between art and other disciplines is also easy to explain: he himself was this rare combination of a scientist, artist, and writer-editor. Like Kepes and Klüver he was convinced that the possibilities for artists to access new technologies needed advancement. In his view, communication between artists about the use of new scientific developments and techniques was just as important, and seriously lacking. Artists should be the first to talk about their work. And in 1968 the printed press was still the main and easiest venue to establish communication on an international level.

Soon new applications of technologies, like holography, or the use of electronics, replaced the articles on the more traditional art forms like painting or sculpture. The shift in emphasis toward interdisciplinary relationships between art, science and technology made *Leonardo* into a forum for exchange between these disciplines and in 1974 the subtitle changed to that effect. The magazine established itself as a professional journal for all those interested in the contemporary visual arts, providing a venue for artists to write about their aims, methods and experiences. It promoted the "interaction of artists, aestheticians, art historians, art teachers, researchers in visual perception, psychologists as well as scientists, engineers and technologists," seeking to bridge the gap between these disciplines and to do so internationally. From this it follows that *Leonardo* never intended to be or become a magazine at the newsstand. Its subject matter was too difficult and specialized - too 'dry' - to read for the larger public. *Leonardo* has followed the developments of the computer and art quite closely over the years. There is, however, one more thing that seems quite important. In reviewing the contributors during these first seven or eight years, one hardly comes across writers from the art-theoretical or art-historical disciplines. The artists' contributions describe mostly the artist's work. The general and theoretical articles have almost always been written by scientists or engineers.

When Frank J. Malina died in 1981, the magazine had to assess its existence once again. Roger F. Malina took over as executive editor (until now). In his editorial about the matter, entitled "Leonardo: a new beginning," he stated the continuation of the previous aims, to be expanded in scope "to match the broader needs of contemporary artists and the science/technology community affiliated with the arts." Roger Malina wanted to include "music, media, performance, language, environmental and conceptual arts - especially as they relate to the visual arts, or technology; art forms hitherto neglected. He also emphasized the magazine as being a "forum for discussions of legal, political and economic aspects."[154]

The International Society for the Arts, Sciences and Technology (ISAST) was founded as an "official network of artists, scienctists and engineers" and *Leonardo* became its official publication. From now on the subtitle was: Journal for the International Society for the Arts, Sciences and Technology. From the Editorial Board: "In many ways analysis was the paradigm of the preceding epoch; however, we believe that our era demands synthesis - searching for common threads in the artistic, scientific and technological aspects of our fragmented cultural fabric."[155]

During the following decade and until today *Leonardo* has maintained this reputation, not only following new developments closely but even trying to be ahead of the times. In the eighties it printed articles on artificial intelligence, the new concepts of order in the natural sciences and the visual arts, fractal geometry, chaos theory and so on. *Leonardo* began to collaborate with other organizations, publishing special issues and supplements for a wider public. Among those were an issue on "Psychology and The Arts," "Jacob Bronowski: A Retrospective," "Visual Art, Sound, Music and Technology," and "Electronic Art," published in association with the First International Symposium on Electronic Art at Utrecht, Netherlands. Together with AMC / SIGGRAPH (the Association for Computing Machinery's Special Interest Group on Computer Graphics) an issue on new developments and applications of the computer in art was published on the occasion of their annual conference in 1989, entitled "Computer Art in Context." The nineties continued with information on the rapidly developing computer technologies, and the communication and information networks. The role of education in respect of new technologies was assessed regularly as well.

Lately, the issues seem, more than in the seventies, to be discussed within the larger social, environmental and political realm. The September 1993 issue presented, for example, the work of ecological artists, like the Harrisons, discussing the social responsibility of the artist. Maybe the fact that *Leonardo* has been inviting special guest editors from different disciplines contributed favorably to its readability. Topics are written with a broader public in mind. Being the magazine it is, and after the publication of the electronic newsletter *FineArt Forum* in 1988, the installation of a site on the World Wide Web could not wait long. The address is: <http://www-mitpress.mit. edu/Leonardo/home. html>

PART II

Chapter 4

Hans Haacke: Systems Artist

Hans Haacke and Jack Burnham: Exchange of Ideas

Hans Haacke is been the only artist who has consciously applied the systems approach both practically and theoretically. The history of his development shows a conscious application of its concepts from a relatively simple level toward an ever more complex layering of the issues researched.

Hans Haacke appears to have played a major role in the development of Burnham's systems ideas. For Jack Burnham, Haacke's work represented a prime example to sustain his ideas of a new "systems esthetics." Hans Haacke and Jack Burnham met in 1962 at Temple University in Philadelphia, when the artist was still working on metal and glass constructions, which Burnham thought were "very optical, cold and precise - interesting and timely - but lacking that spark of sensuous playfulness," and he considered Haacke a "reserved and pitiless technician."[156] He was to revise this opinion soon after he met Haacke. In an interview with Jeanne Siegel, Hans Haacke has said about their early relationship: "Since then we have been in contact and have had a very fruitful exchange of ideas. It was Jack who introduced me to systems analysis..." In the same interview he continued: "Sometime in '65 or '66 I was introduced to the concept of systems. I heard about systems analysis, and the related fields of operational research, cybernetics, etc. The concepts used in these fields seemed to apply to what I had been doing and there was a useful terminology that seemed to describe it more succinctly than the terminology that I and other people had been using until then, so I adopted it. But using a new terminology doesn't mean that the work described has changed. A new term is nothing holy, so it can serve as a union label. On the other hand, a clear terminology can help stimulate thinking."[157] For the next twenty years, Burnham not only followed Haacke's artistic growth, but the two men also developed a friendship in which the writer and the artist were inspired by and drew ideas from one another for their respective theories of art. While Burnham was about to formulate his concept concerning the new directions in sculpture as a systême d'art, Haacke started to work with water, wind, light, heat, cold to visualize certain processes under predetermined conditions. These were the condensation boxes, simple plexiglass boxes in which the viewer could register a continual cycle of water droplets evaporating, condensing, falling, increasing, condensing again, etc. The prevailing cyclical changes were primarily dependent on outside light and heat sources. One can say that the processes as well as the results were interdependent with the environmental conditions present, constituting a "set of relationships in action." Made between 1962 and 1964, they were the first pieces, to use natural phenomena and laws analogous to nature.[158] Wind, air, heat, cold were the artist's main materials in the early sixties, to which he added so-called biological systems during the second half of the sixties. In 1967, he made *Grass Cube,* a cube of clear plexiglass with grass growing in the soil on its top. At the *Earth Art* exhibition in Ithaca, NY (1969), he presented a conical earth mound with a nine feet diameter and approximately three feet high, where he sowed a fast-growing type of winter rye combined with annual rye seeds, titled *Grass Grows.* Haacke envisioned that the rye inside would grow very fast during the exhibition period, and would have completed its life cycle by the time the exhibition closed due to the indoor conditions. Outside he had streched a rope near a waterfall, so that it would collect the water drops sprayed. Winter conditions were to freeze the water of *Spray of Ithaca Falls,* creating a growing layer of icicles, until the layer would become too heavy or the temperature would rise so that it would start melting again. Asked about the meaning behind his pieces, Haacke stated during the symposium that was organized at the occasion of the *Earth Art* exhibition: "I'm more interested in the growth of plants - growth as a phenomenon which is something that is outside the realm of forms, composition etc., and has to do with interaction of forces and interaction of energies and information."[159]

Some of the best of Haacke's efforts were shown outside the gallery during these years. They included his *Rain Tree,* a tree dripping patterns of water; *Sky Line,* a nylon line kept aloft by hundreds of helium filled white balloons; a weather balloon balanced over a jet of air; and a large nylon tent with air pockets designed to remain in balance one foot off the ground. As an art experience Haacke's early systems had a limited temporal life-time. Their function as systems consisted in the exchange of energy with the outside environment, through interaction with the environmental conditions, which were only partially controlled by the artist, and sometimes by the public. In his installation *Photo-Electric Viewer Programmed Coordinate System* (1968) infrared projectors, photo-electric cells or white lightbulbs were switched on and off according to the position of the viewers in the room, creating a "responsive environment" that changed constantly with the number and movement of the visitors.[160]

Referring to Bertalanffy, Haacke has explained his definition of a system: "A system is most generally defined as a grouping of elements subject to a common plan or purpose. These elements or components interact so as to arrive at a joint goal. To separate the elements would be to destroy the system. The term was originally used in the natural sciences for understanding the behaviour of physically interdependent processes. It explained phenomena of directional change, recycling, and equilibrium. I believe the term system should be reserved for sculptures in which a transfer of energy, material, or information occurs, and which do not depend on perceptual interpretation. I use the word 'systems' exclusively for things that are not systems in terms of perception, but are physical, biological, or social entities, which, I believe, are more real than perceptual titillation."[161] Jack Burnham's systems approach as a means of analysis and Hans Haacke's own description and interpretation of his work in the terminology of the systems theory functioned almost as a symbiotic relationship, in which one party nourished the other. On the one hand, the systems approach, with its accompanying language, allowed for the literalist interpretation Haacke was looking for, away from formalist and phenomenological descriptions as well as metaphorical and symbolic connotations. On the other hand, it also implied the notion of context. For Haacke, contextualization meant expanding the art work to include the surrounding spatial aspects, the environment, elements of time, change and process, as well as the social and political component. These not only included research into the function of art and the structure of the art world, but also investigations into general social and economic conditions related to it. It is obvious that a formal description of Haacke's work would not suffice, nor would a metaphorical or symbolic interpretation, only an interpretation connecting all these different aspects as interrelated components. The general systems approach, which was just making its full entree, appeared to be doing just that.

Haacke's Biological Systems

Hans Haacke was born in Cologne in 1936. He studied graphics and painting in Kassel and moved to Paris in 1960 where he stayed for two years and met Yves Klein, among others. He also learned about the activities of G.R.A.V. (Groupe de Recherche d'Art Visuel) and ZERO. Another artist whose work impressed him was Vasillakis Takis, who was working with magnetic fields and electric energy/forces. In 1962, he moved to the United States on a Fulbright scholarship. The following years he participated in numerous kinetic art exhibitions, and exhibits of the groups ZERO and NUL in Europe.[162] The major change of direction occurred in 1963 when Haacke began his first experiments with air currents and fluids. These would lead to his condensation boxes. After 1965, Haacke participated in numerous exhibitions with as many different starting points. One should not interpret this as a sign of ambiguity on Haacke's part, to be interpreted as if he kept shifting positions within the pluriform art scene. Like most artists at the time, he hated his work to be labeled. While his early artworks were connected with the kinetic movement, his later works were categorized as Minimal Art, Arte Povera, Art and Technology, Land or Earth Art, Air Art, Process Art, and Conceptual Art. The cubic plexiglass condensation boxes had indeed something minimalistic. Their objective was obviously completely different. As a matter of fact, he wanted to rid himself of the minimalist stigma that some critics had given his neutral-looking material and forms. As he explained: "A very important difference between the work of minimal sculptors and my work is that they were interested in inertness, whereas I was concerned with change. From the beginning the concept of change has been the ideological basis of my work. All the way down there's absolutely nothing static...nothing that does not change, or instigate real change. Most minimal work disregards change. Things claim to be inert, static, immovably beyond time. But the status quo is an illusion, a dangerous illusion politically."[163]

Thus, while Haacke already perceived his works as composed of a complex set of aspects, or relationships, the organizers of exhibitions generally focused on one particular aspect of the art work fitting the concept of their show, whether it concerned the material or formal aspects, the process or the conceptual element. Hans Haacke also participated in all the important art and technology exhibitions, or was at least invited, although his work is not immediately connected with this movement: *The Machine as Seen at the End of the Mechanical Age; Software; Information; Art & Technology.* He never really collaborated with technicians or scientists, but would ask their advice when necessary, and once admitted that he might have profited from a better technical education. Of course, Haacke featured in exhibitions that had as a starting point the utilization of natural media, such as air and earth, and abstract concepts such as space and idea, and finally the social implications of his work. The between nature and technology was man-made according to him, and he held the opinion that "the functioning of either one can be described by the same conceptual models, and they both obviously follow the same rules of operation. It also seems that the way social organizations behave is not much different. The world does not break up into neat university departments. It is one supersystem with myriad subsystems, each one more or less affected by all the others. If you take a grand view, you can divide the world into three or four categories - the physical, the biological, the social and behavioral - each of them having interrelations with the others at one point or another."[164]

Since the early seventies, Hans Haacke's work took a major turn toward the exploration of social or economic networks and their specific function as mediators of power structures. With *Gallery-Goers' Birthplace and Residence Profile* at the Howard Wise Gallery in New York (November 1969) the focus of Haacke's activities moved toward the investigation of socio-economic systems.[165] Visitors were asked to mark their place of birth and current residence with pins on large-scale maps of New York City and surroundings, with a fifty mile radius. An extended version of the piece, with photographs taken of the locations indicated, was subsequently exhibited at Paul Maenz Gallery in Cologne. The result of the investigation showed that the art-interested gallery-going public was largely confined to certain, mostly rich areas (e.g. the Upper West Side, and the loft district with its artists' residences) in the city. Although not explicitly stated, it was also an indication of the social strata that comprised the art public. However, Haacke professed only to present information structures and to leave it to the viewer to extrapolate conclusions from them. As a consequence, he was repeatedly rejected by major museums, starting with the notorious 'Guggenheim affair' in 1971. The project *Shapolsky et al. Manhattan Real Estate Holdings, a Real Time Social System and Sol Goldman and Alex DiLorenzo Manhattan Real Estate Holdings, a Real Time Social System* consisted of mapping the real estate interests of two holders, Shapolsky and Goldman, in Manhattan, New York in which he showed one to be primarily dealing with slum property, the other to possess the largest real estate conglomerate. As documentation of the proposed project abounds, I will not go into detail here.[166]

Art as an Information System

In 1970, Hans Haacke was asked by Jack Burnham to participate in *Software.* He had computer programmers at the Massachusets Institute of Technology write the program for his proposed *Visitors' Profile.* Each visitor was presented with a number of questions and inquiries as to his or her status, and began the interaction with the computer by typing a self-chosen code identification. The computer compiled the answers, compared them with information received from other visitors, and correlated data relevant for a statistical breakdown. At the exit area of the exhibition, a teletype terminal printed the processed information in the form of statistics. At the same time, a closed-circuit television set-up projected the constantly changing data onto a large screen, so that it was accessible to a great number of people. Based on their own information, gradually a kind of statistical profile of the exhibition's visitors emerged.[167]

By now, Haacke had switched from experimentation with natural systems to a concern with social systems. The *"Polls"* that he presented as art works could still be analyzed in terms of a systems structure. They connected the structure of art (internal) with a social (gallery/museum visitors) structure (external), they were 'works in progress,' coming into being during exhibition time, being as it were created by the visitors/participants as an open system whose result was never reached. However, for *Visitors' Profile* Haacke did no longer use the conventional questionnaires, which were subsequently rearranged into statistics, but he used the new computer data processing possibilities to create what was to be called an information system. Also, the computer network was set up as a real-time system. This expressed Jack Burnham's interest in the ramifications of the computer, which he wanted to exemplify in this exhibition. The significance of this exhibition lay not so much in the fact that it was the first exhibit of its kind. The issue was rather to create a new operational and functional framework for the visual arts, and to move beyond the 'classical' boundaries, into a realm that was just about to develop in full force, that of computer technology, or, more precisely, information data processing technology.

Burnham emphasized the computer's communication and message aspects and their importance for the arts as a device that made possible the change from the work of art as a one-way communication system to a two-way exchange process between observer and work observed. It expressed the idea that the observer could actively participate in the work of art, as we already encountered in Hans Haacke's *Visitors' Polls.* The terminology that Burnham used to describe and interpret this communication system was derived from cybernetics.

The Artist as Social Scientist

Hans Haacke has always been very explicit about the raison d'être of his works as art. They were created as art, to function solely within the art world and were presented within the art context. Even if his methods in many ways resembled those of the social scientist, Haacke himself never pretended to be one. However, he has always abstained from any additional, whether symbolic, metaphorical, aesthetic, or even personal, interpretative commentary on the work, presenting factual information, clad in a form that is identical with the ways data are generally handed to us, an approach that did not allow the viewer to be seduced by what one might call a surplus value, or the aesthetic value of the artwork. For many critics Haacke's visitors' polls and other investigations into the power structures of the art world were difficult to appreciate. In fact, Haacke's writing made it virtually impossible for critics and scholars to circumvent the systems vocabulary in their writing. However, when it came to placing his work within the art context, such criteria as beauty were applied and Haacke was often criticized for not being aesthetically pleasing. Why was this art, if it did not transcend the level of factual information.

Indeed, the presentation of accurate information without further comment or beautification had many opponents, not only among critics, but among organizers and curators as well. In critical reviews in which a comparison was made between Haacke's approach and scientific methods, the artist's work was often judged as being neither one nor the other. There have been few serious attempts to compare Haacke's work with that of the social scientist, but it is worthwhile to refer to one by social scientists Howard Becker and John Walton, because it sheds a light on how his work was analyzed by scientists from another discipline and what kind of criteria they applied. While pretending Hans Haacke's investigations to be social studies into the "network of relationships through which power is exercised in the art world and in the social, economic and political bases of that power," and placing his research within the framework of a social theory, the authors conclude that "even if some of Haacke's procedures appear amateurish or inadequate in the light of some notion of an acceptable social science standard...his results are credible and acceptible by social science standards. Like many good social scientists, he has viewed the phenomena he studies as connected in a system. ... The findings of each individual inquiry both rest on and reinforce the findings of all other works, so that the analysis of the system, considered as a whole, is more credible than any one set of the findings might be."[168]

For the authors, the differences between Haacke's investigations and that of the social scientist finally amounted to the fact that Hans Haacke operates in the world which he investigates, contrary to the scientist, who generally researches problems outside his or her daily environment, sufficiently segregated from it so as not to offer the provocation that Haacke's work evokes. Their assessment that this is the reason why the work creates such reactions is an accurate one. Haacke operates in the art world, not in the world of science, and the purposes of his work are non-scientific. Yet, the problems and issues he tackles are of such complexity, being issues directly related to complex structures of our western society, that they require a very concise methodological approach so as not to appear superficial in the first place. That much is certain: the complexity of the work requires an interpretation that systematically takes into account all related aspects, and this complexity has grown relative to the issues dealt with.

Critical and scholarly interpretations have taken up new positions since the early eighties. A new generation of art writers and scholars has meanwhile come to the fore in the United States. This generation is not only informed about Marxist ideology and the Frankfurt School but has also been influenced by the new French philosophies after Roland Barthes and Claude Lévi-Strauss. Philosophers like Michel Foucault and subsequently Georges Bataille, Jean-Francois Lyotard, and Jean Baudrillard were translated into English, and have been quite influential in academic circles since the beginning of the eighties. In particular, Jean Baudrillard's ideas on the simulacrum have filtered into American aesthetic theories of the visual arts. Their influence on writers like Benjamin Buchloh or Fredric Jameson is apparent and openly admitted.[169] Haacke himself appears to be little influenced by these philosophies. His writing, at least, shows few traces of the so called postmodernist theories. While Haacke perceived his early systems works as an applied form of the systems theory, which was, at the same time, congruous with the scholarly and critical analyses of the work, for both used the same research methods and language, the current interpretation of his installations does not have this 'symbiotic' relationship. This might be due to the fact that the objectives of these philosophies are in a way meta-philosophies. What is discussed is not so much the art or art object itself, but its effect or reception, i.e the consequences in a so-called post-modernist culture.

Continued Controversy

In many respects, Haake's artistic development was timely. He partook in the major shift that took place in the arts of the sixties, described above, which embraced new formal and aesthetic investigations to transgress the boundaries of art. Hans Haacke has continued this research into the cultural significance of the art object within the political, social and economic systems which function as major forces in the institutional structure of the art world. His analyses have become more focused and complex, although his strategies have not changed. And even well into the eighties Hans Haacke's proposals and projects have given occasion to protest by museums and institutions which invited him in the first place. For time and again Haacke has displayed the connections between the worlds of business and art. His installations keep revealing data which are uncomfortable to his hosts, and which they prefer to remain hidden from public knowledge. Multinationals and large corporations like Mobil, Allied Chemical and Philips have been subjected to scrutinous research into their products, markets, and media strategies, into possible connections with military and totalitarian dictatorships, into environmental pollution or toxic waste dumping, etc. Famous collectors like Peter Ludwig, Cologne, and Saatchi & Saatchi, London, have not escaped the artist's watchful eye either. Generally the subject matter is related to the location of the exhibition. Hans Haacke also still writes about the power structures of the museum system. His writing and his art belong together, presenting factual information about how power is exerted and how it functions within the network of social, political and art systems.[170]

The path followed is, in a sense, a continuation of the systems approach as advocated by Jack Burnham, even if theoretically this approach has not been pursued further in art writing and art-historical research. For the various elements which make up the complete work constitute a multi-layered situation that justifies comparison of the works to a system of interacting parts, in opposition to the single independent art object. The relationship of the work to its 'environmental' space implies the introduction of context, making it act as a "set of relations" which all have to be looked at in order to determine its possible meaning, often involving a participatory role of the viewer.

Haacke: "I often develop my works in relation to the specific context in which they are to be exhibited. The social and political environment, the symbolic significance - sometimes also the architectonic character of the site - play an essential role. These are materials for me, like the physical elements I use. ... In the institutions of the art world, as in other branches of the consciousness industry of which it is a part, negotiations take place over how we think about ourselves, about the world, and about our social relations. It is irrelevant whether the negotiating partners are aware of this process and understand what is at stake. The art world is a site of symbolic and as a consequence, also of political power."[171]

Chapter 5

Robert Smithson: In Search of a New Paradigm for Sculptural Space

Theoretician of the Earth Art Movement

Robert Smithson's role in art, both as artist and theoretician, has been identified with the Earth Art movement. His position within it has been analyzed extensively, as it was quickly recognized and co-opted by the art world as something different on the horizon.

Critical interpretations of the movement moved between depreciatory attacks on its ambivalence in striving for a true change of the art system and its intrusion upon the environment to its echoing the growing interest in world ecology.[172] Smithson's own writing has contributed much to a critical and scholarly assessment of his work in a larger context, and in this respect (only), his position was similar to Hans Haacke's. Only Lawrence Alloway initially put forward the possibility of a systems approach in his art, albeit in general terms. Alloway followed Smithson's artistic career closely, and in his excellent article "Robert Smithson's Development" he analyzed the systems aspect of the artist's work as follows: "The translation of a concept of a work of art is typical of Smithson's interest in the relationships of art and the world as opposed to an art isolated by its internal relationships. ... Site and nonsite constitute a collection of relationships among variables. The site is identified by information supplied by the artist in the form of maps, photographs, analogical subjects (bins and trays cued by the original lay of the land), rock samples and verbal captions. The nonsite, by this accumulation of references, acts as signifier of the absent site."[173]

During the seventies Smithson's work was mainly discussed in terms of his contribution to the Earth Art movement, both as artist and theoretician, in which his wide influence was recognized but only vaguely defined. One may speculate whether his premature death in 1973, while flying over *Amarillo Ramp* (a work in progress) did not contribute to the Smithson myth. However, not until Robert Hobbs' publication in 1981, in which he set out to analyze a number of works as representations of the entropy law as Smithson interpreted it, was there a close reading of the thought-out layering of materials, location, and its environmental history.[174]

Robert Smithson had his first exhibition of paintings in 1959 in New York City, where he lived until his death in 1973. The first artists who brought natural materials into the exhibition space - Michael Heizer, Dennis Oppenheim, Walter De Maria, Hans Haacke, Robert Morris, among others, and also Robert Smithson - initially sought to express their discontent with the existing state-of-the-art system. Smithson's writings and statements leave no doubt about his intentions.[175] Between 1968 and 1973, his work changed dramatically, from minimalist objects toward a use of materials which was characterized by him as "de-creating, de-naturalizing, de-differentiating, de-composing, and re-assembling." From there, he moved into the environment altogether, while the formal elements were increasingly simplified, although growing ever more complex in content. The experimentation with new artistic means to expand art's boundaries became a search for a new function of art and a different role for the artist. Although his premature death prevented the further continuation of his ideas concerning land reclamation, herein lies a major distinction between him and his colleague earth artists, none of whom have pursued the consequences moving outdoors entailed. If we accept Hobbs' approach, it would mean that, even if Smithson still stands firmly with one foot in Earth Art, his work could also be interpreted as being of an "organized complexity" that evolved from this search for a new context for art.

Smithson's critical reviews are an early clue. Not only did he suggest another type of art analysis, which would involve an interdisciplinary approach, in which the art object is perceived in a broader context, beyond a linear chronological time model, but he also noted the relevance of language as a principal mode of communication, whereby the syntax of art became an aesthetic mode of communication. The ideas of art as communication and art as information were relatively new at the time, and had entered the art discussion via linguistic theories, which in turn derived some of their ideas from the information and communication theories.[176] In the 1940s, the word information had gained new scientific meaning. The concept of information was defined mathematically to satisfy the needs of engineers and communication experts. It was Claude Shannon from Bell Telephone Laboratories who had suggested a relationship between the laws of entropy, of energy and information in a mathematical equation (1948).[177] That Smithson was indeed familiar with these concepts is certain, since he referred to them in an interview later on, in which he said: "In information theory you have another kind of entropy. The more information you have the higher the degree of entropy, so that one piece of information tends to cancel out the other. The economist Nicolas Georgescu-Roegen has gone so far as to say that the Second Law of Thermodynamics is not only a physical law but linked to economics. ... One might say the whole energy crisis is a form of entropy."[178]

Another source was George Kubler's *The Shape of Time - Remarks on the History of Things* (1962), a book highly popular with artists at the time. Apparently Kubler's book was on the shelves of virtually every artist. For him both stylistic and iconographical approaches were too limited to explain the developments in art. So was the individual biography. He also denounced the use of biological and physical metaphors, common in art-historical language, as not really appropriate, and suggested that a terminology derived from physics might be better suited.[179] The separation of art and science was obsolete for Kubler, in that this separation was based on a distinction between the fine arts and the applied arts, which was no longer functional, whose consequence was that similar processes in art and science had not been looked at from the same historical perspective. George Kubler perceived art as part of the history of things. Kubler's proposition included both artifacts and art works, both unique objects and replicas, tools as well as ideas and literary means of expression, all materials made by human hand, ruled by related concepts that were developed over a certain period of time. He was convinced that one could read a "shape of time" from the totality of all these forms. Thus, he thought to move beyond the art-historical treatment of the visual arts as a typical Western phenomenon, to define the intercultural aspects. Instead of developments separated from each other by well-defined time-bound stylistic non-functional characteristics, he proposed to reformulate the notion of art to comprise all artifacts - as things - made by mankind, whenever or wherever.

This is an important distinction, because intercultural characteristics are not necessarily bound to a chronological linear periodicity, since similar phenomena can reappear at different times in different places. George Kubler was also interested in the new ideas that were introduced by linguistic structural analysis, which in turn was influenced by current ideas concerning the notion of the message in information and communication theories. For example, he discussed art in terms of a signal, transmitted by a sender and received by a receiver. Because a signal is always received with delay, a relais, there is a certain amount of deformation in the way the signal is finally received as information. Certain is that Kubler's ideas about art, in opposing the prevalent definition of art and existing art-historical methods, in proposing an intercultural and interdisciplinary approach, although general, attracted those artists who wanted to leave the existing boundaries of art. For Kubler, our material innovations were all things, whether they were art works, artifacts, or man-made environments, which could be subjected to the same interpretation. He asked himself: what happens to these things in the course of time, what kind of value system do we apply to things from the past? These thoughts also interested artists working to create non-objects, non-precious works of art.

Art as Entropic Phenomenon

Robert Smithson's writings show the artist's diverse interests, ranging from the earth sciences, geology, and crystallography to science fiction and experimental film. After writing a number of critical papers, his writing soon revolved around his own art, in order to evaluate and clarify the issues he was concerned with. Smithson's essays sometimes appear like fiction, because he often went off at multiple tangents, and connected issues that on the surface did not seem to relate at all. One finds this new preference for non-linearity also in the writing of Marshall McLuhan, for example. Of course, the idea that things do not always move linearly from a to b in a cause-and-effect sequence, but that there are often multiple variables affecting the outcome, had become the accepted approach in systems analysis and other sciences. One may suspect that this approach put a stamp on Smithson's mode of working.[180] I will not speculate how deep the knowledge of the literature he referred to was, since Smithson had a tendency to drop names without specific reference to his sources, but he studied the geological sciences, crystallography, and related disciplines thoroughly, and was well aware of the latest developments.

A much discussed topic in the fields Smithson was interested in were the workings of the entropy law. It is fair to say that the notion of entropy runs as a major thread throughout Robert Smithson's artistic oeuvre. His writing is permeated with the concepts and vocabulary of this law.[181] It confirms his familiarity with the concepts of entropy and the second law of thermodynamics, current at the time as well as their relationship to the new thoughts that were being developed in related sciences such as natural physics. Entropy had become associated with the "running down of the universe," with an ever-increasing loss of energy, a condition which would ultimately lead to its heat death. Smithson did not share the fatalistic conclusion that the earth was nearing its end, but he did share the point of view that, everything being entropic, meaning subject to entropy, "one's mind and the earth are in a constant state of erosion." For him nature was for ever developing and changing, never finished, so that "if the work has sufficient physicality, any kind of natural change would tend to enhance the work. Geology has its own kind of entropy, that has to do with sediment mixtures. Sediment plays a part in my work. Unlike Buckminster Fuller, I'm interested in collaborating with entropy. Someday I would like to compile all the different entropies. All the classifications would lose their grid."[182]

Smithson introduced the term entropy in his first major article, "Entropy and the New Monuments" (1966), as a sort of critical metaphor for the condition of Minimal Art forms. According to the artist, these "monuments" were symbolic for entropy at work. Made of artificial materials, such as plastics, chrome, light, etc., instead of natural materials, such as marble, granite or other kinds of rock, these works, according to Smithson, were involved in a systematic reduction of time down to fractions of seconds, rather than in representing the long spaces of centuries. He held the opinion that these artists had created an art that could be interpreted as a visible analog for the second law of thermodynamics, which stipulates that in the "ultimate future the whole universe will burn out and be transformed into an all encompassing sameness."[183] It is not relevant whether Smithson's interpretation is right or wrong. Relevant is that the artist ascribed to art the same entropic properties as to everything else on earth, and that this approach allowed him to perceive art in a new contextual mode: since art was subject to the law of entropy, which made it a function of time and space, it should be perceived as dynamic instead of motionless. This served as a starting point and a theoretical foundation of his future works. Unfortunately, the notion of entropy has turned out to be one of the most confusing and misconstrued concepts, and in terms of

Smithson's art one has to look at the knowledge he could have had and the ideas that were current at that time.

Smithson's sculptures of these years (1966-1967) almost always consisted of geometrical arrangements and superficially appeared minimalistic. However, their mathematical proportions relied on crystalline forms, such as the hexagon. Ice crystals are such hexagonals, and in terms of entropy, one might say that ice can be seen as water's entropic state: energy in its most reduced state. Smithson's *Alogons*, made in 1966, can probably already be read as testimonies of entropy. For the understanding of these works it is important to recall that in crystallography the concepts of entropy were and still are highly relevant.[184] The study of entropy, which is, among other things, concerned with energy transformation and transmission of information, implied the study of concepts already introduced: open and closed systems, randomness and disorder versus order and geometry, indeterminacy and uncertainty versus determinacy and certainty, and time and space. As said before, in contrast to the so-called mechanistic world view, all these aspects are now considered interrelated and interdependent. Therefore these words were introduced into art writing at the time to describe the kind of work that deals with notions of growth and decay, hence exchange of energy.

In order to systematize his view, Smithson developed the theory for which he became most known: the Site/Nonsite dialectic. It dealt with these aspects in pairs, two variables functioning as each other's opposites.[185] Smithson's intent had always been to move away from the formal criticism of the sixties, which looks at the object as an isolated thing. The Site/Nonsite dialectic indeed appeared to have all the characteristics of a totally new direction in art. With the Site/Nonsite idea, he had created a physical and conceptual set of relationships that he visualized with works that were partially outside (site-related), partially inside (room, gallery related). This is particularly apparent in a series of mirror and glass works, called *Strata* (1968-1969). The glass pieces were arranged according to geometric principles, and looked highly ordered, and structured. Contrary to its appearance, though, the non-visual molecular structure of glass is random, based on a non-order principle, while the molecular structure of crystals is regular. This contrast is dealt with in many books on crystallography, so that we may assume that Smithson made a reference to the dichotomy between the visual and non-visual aspects here, i.e. to the distinction between appearance and reality.[186] A close reading of some of the mirror and mica works, made also in 1968 and 1969, brings to the fore the same concerns. For example, *Untitled* (mica and glass, destroyed) consisted of alternating layers of glass and mica, whereby the artist created a visual structure that showed a perfect regularity in the composition of the glass sheets, alternated with random looking mica pieces. Actually, the molecular structure of glass is neither crystalline nor regular, and the seemingly irregular mica has a neat molecular arrangement. Again, the artist played upon the contradictions that often exist between the external world and its internal composition Smithson himself referred to the discrepancies that can exist between the reality of the visible and the invisible in his article "Entropy and the New Monuments."

The same concepts underlay the *Cayuga Salt Mine Project*, defined as a Site/Nonsite project, conceived for the *Earth Art* exhibition in Ithaca, New York 1969. This *Mirror Displacement* originally was to connect the Cayuga Salt Mine with the indoor museum space. In the mine Smithson had placed eight mirrors. Another eight mirrors were evenly spaced between the mine and the exhibition space, called *Mirror Trail*. Salt of the mine was brought into the gallery. On the walls there was a display of geological maps of the area, photographs of the mirrors in the mine and those of the trail to the museum. Here the salt served as a container for the mirrors, instead of the containers holding the material. Again, Smithson played on the material and formal contradictions that existed between the visible and the invisible, between order and disorder, between reality and appearance. Although salt has an amorphous appearance, it has a regular molecular structure. Glass, on the other hand, looks regular but does not have a regular molecular arrangement. Elements of crystalline structures are applied to a system of mapping that connects the outdoor space with the indoor space and artwork. Although concepts of space and scale, time and history, perceptions of a two-dimensional and a three-dimensional reality, are treated as related opposites, they also constitute a complex of layers that may be compared to a network whose extensions are interdependent. These extensions did not remain within the boundaries of the work itself but were extended to outside territory, i.e. context. The terrain could be the actual site where he placed a mark, photographed it, and subsequently removed it. It could be the dialogue between an interior and exterior site. It could be the site at large, its geographical, geological, historical, economic setting, and so forth.

A simple description and interpretation of the visual structure of the works described above would not have explained much. The non-visual references to crystalline structures and the implications of entropy are equally important. As explained, entropy is a mathematical expression describing disorder. To put it differently, it is a function of the organization of a system. Thus, one may deduce that the artist conscientiously thought out all these variables, connecting them with the idea of ordering which art also is. Smithson himself has explained his view in general words: "An artist in a sense does not differentiate experience into objects. Everything is a *field* or a maze, and you get that maze, serially, in the salt mine in that one goes from point to point. The seriality bifurcates: some paths go somewhere, some don't. You just follow and what you're left with is like a *network* or a series of points and then these points can be built into *conceptual structures*... I am using a mirror because the mirror in a sense is both the physical mirror and the reflection: the mirror as a concept and abstraction; then the mirror as a fact within the mirror of the concept. So that's a departure from the other kind of contained, scattering idea. But still the bipolar unity between the two places is kept. Here the site/nonsite becomes encompassed by the mirror as a concept - mirroring, the mirror being a dialectic. ... The route to the site is very *indeterminate*. It's important because it's an abyss between the abstraction and the site; a kind of oblivion... A trail is more of a physical thing. These are all variables, indeterminate elements which will attempt to determine the route from the museum to the mine. I'll designate points on a line and stabilize the chaos between the two points. ... There's no order outside the order of the material. I don't think you can escape the primacy of the rectangle. I always see myself thrown back to the rectangle. That's where my things don't offer any kind of freedom in terms of endless vistas or infinite possibilities. There's no exit, no road to utopia, no great beyond in terms of exhibition space. I see it as an inevitability; of going toward the fringes, toward the broken, the entropic" (italics my emphasis).[187] As said before, the key terminology used by Smithson to circumscribe the direction he was moving in, common in physics, mathematics and systems theory. This means that Smithson recognized similar systems at work in disciplines as different as crystallography, cartography and art.[188]

In his most renowned work, the *Spiral Jetty (*Rozel Point, Great Salt Lake, Utah, 1970), the time-space continuum is introduced in all its entropic ramifications. The spiral itself, being the dominant formal element, implies this. Apart from the many cosmic and symbolic meanings of the spiral, which are certainly referred to, in the context of Robert Smithson's work the meaning of the spiralling form should first be sought in the realm of crystallography. The artist had already used it in an earlier piece called *Gyrostasis*.[189] The concept of gyrostasis refers to a mode of crystal extension, known as screw dislocation, a geometric mathematical ordering of the hexagonals that eventually spiral around a void. This means that there is no single vanishing point. Since the Renaissance, the universal focal point had been accepted as the three-dimensional rendering of spatial representation, as *the* representation of reality. The spiralling form included the notion of a "pointless vanishing point," which became for Smithson a valid method to visualize a reality hitherto invisible, but just as real. This in itself would not be that interesting, had it not been related to the idea to break through the known toward a new way of seeing, whereby the old concepts of absolute space (and time) were left behind. For Smithson it implied a new vision of reality: the reality as we view it is anthropocentric, it is a mental construct, a conceptual framework. It was on this level that his art and the scientific concepts he applied were connected.[190] The *Spiral Jetty* is an art work that functions as a system of interlocking elements. It can only be understood if one takes into account all the different aspects of the surrounding environment - the external conditions - as well as the invisible structures, and finally the spectator him- or herself. The spiral turns left instead of right, and is asymmetrical: the salt crystals that grow on the banks of the piece have an asymmetrical molecular structure. Also, salt crystals are said to have some connection with 'being' and 'matter,' in that they may assume a living as well as a crystalline appearance. Salt Lake is a dead sea, it has reached it most entropic state, as Smithson would say. The entropy law is the force that governs this work. Time, space, the macroscopic and microscopic, the living and dead elements, together are all interrelated elements of the *Spiral Jetty*. Finally, the work is constantly subject to change: it grows and erodes; when the water level of the sea rises, it disappears under water. *Spiral Jetty* was under water for a while indeed, and has recently reappeared above the surface. These changes, however, are only partially determinate. Smithson foresaw the kind of change as part of the earth's entropic processes, but how exactly they would take place was left to and ruled by chance and indeterminacy. To experience the work the viewer had to 'enter' it and follow the spiral track to the end and return. Although the experience of the public seems not to have been much on Smithson's mind, viewer participation became an important aspect of much outdoor sculpture. The development of the piece and its construction, as well as the making of the film about the building of the work, are described in detail by Smithson himself in an article.[191] These concerns were far removed from those of Minimalism which was principally interested in reducing form to its physical essence, as well as the initial Earth Art ideas, rather being related to those concepts in the sciences that dealt with an interpretation of things, whether organic or inorganic, whether part of an open or closed system, as sets of relationships in action.

Concluding, one can say that entropy is made visible in his art and writing as a major force that governs all matter and living things on earth. Since the entropy law is not an isolated phenomenon, but connects time and thus history, its characteristics offered Smithson the possibility to use this concept to develop a mode of art that has to be perceived in terms of relationships. Smithson himself has said so much: "Looking at the nature of the park [Central Park], or its history and our perceptions of it, we are first presented with an endless maze of relations and interconnections, in which nothing remains what or where it is, as a-thing-itself, but the whole park changes like day and night, in and out, dark and light - a carefully designed clump of bushes can also be a mugger's hideout. The reason the potential dialectic inherent in the picturesque broke down was because the natural processes were viewed in isolation as so many classifications, detached from physical interconnection, and finally replaced by mental representations of a finished absolute ideal."[192]

Smithson's Attitude Toward the Land

Until 1969, Smithson had made only temporary outdoor works, most of them called mirror displacements, which he had made during his travels. The photographs remained the only witnesses of his activities. One of the best-known of these displacements was *Incidents of Mirror-travel in the Yucatan*, on which Smithson published an essay and photo reportage in *Artforum* (September 1969). When Smithson was informed that his proposals for the *Art and Technology Program* in

Los Angeles had been refused, he gradually started to think about other possibilities to work outdoors. From ca. 1970 onward, he did a number of permanent works in the environment that were characterized as 'pour' works. Asphalt, cement, glue, tar were poured onto and into an industrial, generally already run-down location. The processes, the running of the materials, the drying, and the subsequent deterioration and weathering were all part of the piece (*Asphalt Rundown,* Rome, October 1969; *Concrete Pour,* Chicago, November 1969; *Glue Pour,* Vancouver, 1970; followed by the more well-known flow/pour piece *Partially Buried Woodshed,* Kent, Ohio, January 1970). Compared to other artists working with randomized processes, one may distinguish Smithson's work by the fact that the nature of the chosen material was connected to the history of the site, in particular its geology, as well as the human interventions over a period of time. Again, location and material, as well as the activity itself, connect the works with his vision of art as subject to entropy. They prefigured the *Spiral Jetty.*

While these works have tied Robert Smithson firmly to the Earth Art movement, he himself was obviously looking for another context. This is corroborated by the ideas on land reclamation that he had begun to develop in the early seventies. This aspect of his activities has not been given much attention, probably because none of the proposals resulted in a tangible work. Earlier, Smithson had already offered his services as artist-consultant to an architects/engineers firm. In "Towards the Development of an Air-Terminal Site," he described his position and the process he became involved in, "not as an architect or engineer, but simply as an artist. The discussions do not operate on any presupposed notion of art, engineering or architecture."[193] Although the project ultimately fell through, it instigated Smithson to develop some of his ideas on the possible function of an art outside the gallery and museum system, when he envisioned that the artist could play an important role in reshaping the earth, in reclaiming a landscape that had been drained of its energy, devastated and ruined by the methods used by (strip) mining companies.[194] In 1971, soon after he had completed *Broken Circle/Spiral Hill* (*Sonsbeek 71*, Arnhem, NL), he drafted his first land reclamation statement in which he called for the recycling of natural resources through Earth Art: "Across the country there are many mining areas, disused quarries, and polluted lakes and rivers. One practical solution for the utilization of such devastated places would be land and water re-cycling in terms of 'earth art'... a dialectic between land reclamation and mining usage must be established. The artist and the miner must become conscious of themselves as natural agents. Economics, when abstracted from the world, is blind to natural processes. Art can become a resource that mediates between the ecologist and the industrialist. ... Art can help to provide the needed dialectic between them. A lesson can be learned from the Indian cliff dwellings and earth works mounds. Here we see nature and necessity in consort."[195]

Smithson's correspondence proves that he spent most of 1972 in contacting mining companies and such to find corporate support for his land reclamation ideas.[196] Only a commission from Charles Melby of Minerals Engineering Company and a proposal *Lake Crescents - Forest Park South* received promises of funding, although the timing seemed perfect, as other environmentalists and those who fought for the ecological cause were receiving favorable public attention. But Robert Hobbs has suggested that the corporations concerned preferred to maintain a low profile in the midst of the political battle. Also, Smithson's proposals may not have been quitewhat they had in mind themselves. A brochure of Hanna Coal Company "decrees that the land will be restored to its maximum usefulness - a flourishing area, treed and grassy, providing income from cattle and lumber, recreational lakes, even ski runs." Answered Hobbs correctly in my view: "Smithson might well have asked himself, as the managers of Hanna Coal may have asked themselves: where is the need for Earth Art in the recreational paradise Hanna is promising?"[197]

When Robert Smithson moved his activities outdoors, away from the gallery system, he expanded the terrain of art not only physically. He moved into other systems on organizational and institutional, economic and political levels. From the sole painter or sculptor in his or her studio, the artist now had to work on a team. As artist-consultant, he worked with architects and engineers, among others. The construction of the *Spiral Jetty* became similar to the execution of a building project: it involved setting up contracts for the lease of the land, finding the financing, hiring the construction crews, supervising the construction, handling publicity, and so forth. Collaborating with other disciplines also meant that one had to be open to other people's advice or ideas. In addition, he became a developer, manager, and supervisor. That the economics involved in this type of project and environmental art were also of a completely different scale, is obvious. Smithson did not necessarily intend to assume the position of a manager or contractor. Yet the nature of this new type of art required a different role for the artist, a role which has been accepted by the 'second generation' of artists working in the field of public art.

Ecology and Technology

Smithson did not belong to those who held science and technology to be the solution to the world's problems, bringing welfare and happiness to all, but neither was he one of those who only saw Satan at work here. And we should not forget that his earthworks could not have been made without technological research and the necessary equipment. But technological means were tools for him, instrumental in pursuing his projects. He appears to have been rather sceptical about technological progress and so forth, but he accepted the conditions it created, and proposed that it might be better to deal with it in terms of land reclamation. Anything that reeked of officialdom, or came near 'official religion,' turned him off. Robert Smithson showed an utmost dislike toward transcendentalism, the utopian, basically any ideology that seemed to overlook the real problems. He thought that the ideals of the ecology movement, to return to Mother Earth, were basically archaic, and their solutions to problems not based on reality. "... Not that I am opposed to the ecology movement - far from it. One of the things that interests me most, in fact, is the idea of using abandoned quarries, old strip mines, and such places as sites for earth art. These ruined landscapes could be recycled, too, and given over to a different type of cultivation," he said.[198] Paradoxically, he had great admiration for Frederick Olmsted, co-designer of New York's Central Park. Olmsted was an example to Smithson, for he saw (in Smithson's interpretation) nature in terms of real land, recognizing its physicality, its "unexpected conditions" which are subject to "chance and change in the material order of nature." His relationship to the natural environment and the technological landscape were the two sides of a coin. Smithson had a preference for used, industrial materials and deserted wasteland, as prime examples of the entropy law. An avid reader of science-fiction, could this interest perhaps have been stimulated by the gloomy and doomsday-like atmosphere of a destroyed Planet Earth so often portrayed?

Chapter 6

Expanded Moves into the Environment

Next Steps

Smithson's projects and ideas stood at the intersection of a road that was also taken by a 'group' of artists after him. At first, following the wave of large earthworks, many artists went outside to create smaller temporary works. Galleries and museums leapt onto the bandwagon and the number of exhibitions where artists were invited to create a temporary work especially for the site - the museum grounds, or college and university campuses and city parks - grew enormously since early shows like *Earth Art* (Ithaca, NY 1969) or *When Attitudes Become Form* (Bern, Switzerland 1969), and since 1975, have become too numerous to list. *Sonsbeek* (Arnhem, Netherlands 1970-71) was somewhat of an exception, because artists were allowed to make permanent works in the land. So, Robert Smithson's twin *Spiral Hill/Broken Circle* (Emmen) and Robert Morris' *Observatory* (rebuilt Oostelijk Flevoland, Netherlands) still exist and have become landmarks in their kind. The latter still has the appearance of a large earthwork sculpture, and is not truly related to a specific site. It consists of two earth rings with five sections, two 'canals' and four notches, two of which are lined up with V-shaped notches in the outer ring, and oriented to the sunrise at the summer and winter solstices. In order to experience this, one would have to visit the site at those times. Concerning its historical references to ancient observatories, Morris said: "The overall experience of my work derives from Neolithic and Oriental architectural complexes. Enclosures, courts, ways, sightlines, varying grades, etc., assert that the work provides a physical experience for the mobile human body. ... The work's temporal focus - the marking of the four annual sunrises or seasonal changes - moves it beyond a simple, decorative spatial structure."[199] Maybe one can consider these works to be among the first to initiate the development of the new public sculpture.

Soon, the term Environmental Art shifted in meaning to include any activity that was ambient. Other definitions followed, such as art in the land, public sculpture, public art, art in public places, and so forth. From the early seventies onward, the large earthworks were exchanged for another kind of involvement with natural resources and growth processes on the land - whether it was the small back yard, or the little patch next to the front door. Often these activities resembled small-scale farming or gardening projects, or personal and private rituals in nature. California artists experimented in the deserts. Others, like Charles Simonds, chose the urban context. Especially woman artists used natural elements and growth processes as a means of identification, as a search for a new connection of the body with the earth. For example, some of Michelle Stuart's outdoor works of that period resemble a configuration of found stones, and are unobtrusive actions. Few projects, however, reached beyond the search for the self, important as they were for the development of the feminist movement and the new self-consciousness of the woman artist.[200] An example that involved both environmental and social aspects was *The Farm*, also called *Crossroads Community*, located on several acres of unused land in the midst of a freeway intersection and founded in 1974 by Bonnie Sherk with the intention of creating an urban farm model to reconnect city-dwellers to nature. It became a center for numerous activities, educational, cultural and political, and functioned until 1984. Sherk: "The Farm is a social art work. I think of it as a life-scale environmental performance sculpture with a layering of meanings, metaphors, and actual situations. I see it as art, as well as incorporating the divergent fields of all the arts and literature, education, community service, public health, the environment, city planning, local politics, and real estate."[201]

Generally, though, the forms that were chosen do not seem all that new now, calling up memories of ancient architectures, such as celestial observatories, or places like Stonehenge. There are many works that use geometric arrangements of circles and spirals. The maze is another form that artists felt attracted to.[202] The performances were often ritual in character.

The work of yet another 'group' of artists initiated the development of a new kind of public sculpture, or, to use Robert Irwin's terminology, Art in Public Places. These artists began working in and with the environment in order to effect change. Change in terms of a new function of art as a part of the environment, and of its maker as someone who participates in the real world. But not only that. These artists also sought to effect a change in attitude toward nature or society at large, to effect a different mode of thinking. Thus, Earth Art's premises to break with the boundaries of art by utilizing land in far-away places were left for concerns more true to the surrounding environment and our relationship to it; for concerns more true to ecological, social and political implications. It implied new ways of production and presentation as well as reception.

By now quite a few artists have chosen this direction, generically called public sculpture. In order to show its evolution, which included the expansion into other disciplines, into the fields of science and technology, beyond the territory of art itself, the following chapters will discuss the work of a number of artists whose artistic career began at that time and who have continued to work and research in this direction: Helen and Newton Harrison, Nancy Holt, Robert Irwin, James Turrell, and Alan Sonfist. By proposing a more operational approach in the interpretation of this development and its consequences for art and the role of the artist, the focus will be on the contextualization of the work.

JAMES TURRELL's Vision of the Sky

Perceptual Spaces

The *Roden Crater* project is in process of becoming the second most renowned contemporary 'earthwork' in the Western hemisphere after Robert Smithson's *Spiral Jetty*. Seldom has a project received so much attention before completion, through exhibitions of drawings, models and maps, as well as publications in art magazines and news media. Located not too far from Flagstaff, in Arizona, it is easier to reach than the earlier earthworks, but it is far enough from 'civilization' to make the visit a true effort. In 1974, Turrell began searching for a site that met the requirements that he envisioned. A pilot, he flew over almost all of the western regions of the United States, from the western Rocky Mountains to the southern desert areas, before chosing the site of the Roden Crater. In 1977, Turrell finally purchased the crater, and in 1979 constructions on the project could begin. It is scheduled to be completed sometime in the 1990s. Funding for a major project like this has come from the DIA Art Foundation, as well as public (NEA) and private funds. Turrell himself has contributed to the financing through the sales of drawings, photographs and maps. The artist says that he initially conceived of the idea which would become the Roden Crater project in 1972, and he has described the project as an extension of his concerns with light in the late '60s. It has become, as it were, his life-work, growing larger and larger in size and complexity over the years.

Los Angeles-born artist James Turrell became known through his ambient light installations in the late sixties and early seventies.[203] In order to get away from the thing-in-itself, Turrell began to use light as a means to create a spatial environment without visible object but which nevertheless had a palpable physical presence. Using projected or fluorescent light sources and beams, he projected an 'image' on the wall that, from a distance, appeared to have a three-dimensional, almost physical quality. Only when a person would come approach the wall, the 'image' would dissolve into 'light on the wall.' These *Projection Pieces* (1966-1968) developed into works that involved the architectural space, surrounding the viewer to a larger extent. In subsequent pieces, entitled *Shallow Space Constructions*, the artist had a second wall constructed in front of the rear wall, in such a way that some space was left on all four sides. By lighting the wall from behind, it seemed to visually cause the partition wall to float (1968-1969).

In 1968, Maurice Tuchman invited James Turrell to participate in the *Art & Technology Program* (1967-1971), together with artist Robert Irwin and Ed Wortz, an experimental psychologist, who headed the Life Sciences Department at Garrett Airesearch. Robert Irwin, who was still painting at the time, was involved with perceptual psychology rather intuitively, and Turrell, who had considerable academic training in psychology at Pomona College, brought an understanding of the experimental methodology to the project. Wortz' research was concerned with human perceptual response in such special conditions as space travel and astronauts' moon walks. The Garrett Life Sciences Department had been involved in developing life support systems for manned lunar flights. The *Report on the Art and Technology Program* gives a detailed account of the development of the project. Turrell, Irwin and Wortz began their collaboration with a series of experimental explorations on the nature of perception. Their research included a person's reaction to isolation in a completely dark anechoic chamber for different periods of time, the relationships between tone and color perception, the principal aspect of which was the exploration of ganzfields, and the area of alpha conditioning. Turrell resigned from the project in August 1969. Irwin and Wortz continued their collaboration, which resulted in a symposium as their contribution to the *Art & Technology Program.* Called the First National Symposium on Habitability for Long Term Space Travel and Space Stations, the symposium, sponsored by NASA, was concerned with habitability in its broadest sense, and included even speculations on habitability in space and design of extraterrestrial structures.[204]

However, Turrell's research in ganzfields continued as a consequence of these experiments. The result was a series of "Space Division Constructions," like *Avaar* (1976), and "Ganzfield Pieces", like *The City of Arhirit.* The latter consisted of four consecutive spaces made for the Stedelijk Museum in Amsterdam in 1976.[205] A ganzfield is a space in which there is no perception of form due to a homogenous color field. For these spaces, a room was divided into two parts by a wall, in which a rectangular space was left open. No direct light could enter the 'sensing space' from outside. The only light entering the room was reflected off the walls in the room outside, creating an evenly distributed ambient light, which was very homogenous and looked like a green-gray to gray-blue mist. It had the appearance of a kind of semi-transparent surface, that slowly yielded when the viewer approached the wall. Only when one had come very near, another room revealed itself behind the 'surface,' although one would not be able to determine its depth or measurements. In the *Ganzfield Pieces* the visitor is in the 'sensing space.'

The *Roden Crater*

The *Roden Crater* project continues this kind of research on a large environmental scale, to make visible the manifold phenomena occurring in the sky without us being aware them of under normal conditions. The overall theme of the work is to enhance awareness of the perceptual qualities of the light present, day and night, through annual and other timetables. In order to reveal these aspects Turrell relies on whole systems of interdependent phenomena in the sky and on the earth, i.e. characteristics of the crater itself and the immediate surroundings, as influenced by the changing daily, seasonal and annual cycles; the inherent properties of the sky and the qualities of its light, and how they may change randomly under the influence of unpredictable weather conditions, or through erratic patterns in the eruptions of the sun, or the stars. The scientific data have to be brought together and related to one another, to correctly construct the spaces in such a way that they manipulate the light, so that the viewer-participant experiences such ephemeral events as the aurora borealis. The aurora borealis is a reddish glow of light in the northern sky and visible only during intense atmospheric sunstorms that occur shortly after large solar flares, as a result of the interaction between the earth's atmosphere and the sun's atmosphere.[206]

The crater bowl itself will focus the viewer's attention on the phenomenon of celestial vaulting. There will be a series of 11 interconnected chambers, each with a specific perceptual quality, for example, a ganzfield. In addition, the spaces are aligned so that some astronomical events, like the winter or summer solstices, can be experienced once or twice a year. For example, during the winter solstice, the light will shine through three chambers with the sunlight projected straight on the back wall of the third space. The alignment of the moon has also been calculated, but will happen only once every 18.61 years, which is the period of the moon's regression cycle. Upon completion, the *Roden Crater* will disclose a multiple range of perceptual experiences in time and space. Even visiting the site on different occasions will give the spectator different experiences, not only because the light conditions have shifted, but also because the physical and mental dispositions are changed, affecting perceptual experience.

Turrell has described this work as "working with light in a space so that the light feels to be tangibly present there and seems to be filling the space. In *Roden Crater*, I am working with a series of spaces that select light from different portions of the sky. The effects of this light will create certain 'atmospheres' in the spaces inside the crater. These areas inside the crater will react to the light from the sky in certain ways during the night. So there are two aspects, day and night aspects. My light sources are limited to the sun and the moon and starlight, and of course, reflections of materials like sand or snow or things that vary with the season. I involve the light in the space with imaged events, and these events are like a syncopated rhythm that comes from the heavens."[207] Another site-specific reference, mentioned by Turrell, are the kivas of the neighboring Hopi indians. Kivas are underground holy spaces. They function as ritual cult sites.

When the artist speaks about his work in this manner, there appears to be somewhat of a dichotomy between the purely scientific research involved to obtain the desired results and the emphatic emphasis he places on the intuitive, even emotional, experiences in the approach. Certainly the work will not serve scientific purposes; the *Roden Crater* is not science, the artist insists. It is a work of art that is powered by the movement of sun, moon and starlight. Its function is to be an art work. However, his methods are to a large extent scientific and the artist could not have progressed without the scientific and technological research and assistance provided by astronomy, geology, topology, climatology, and so forth. The artist is advised on the astronomical implications by E.C. Krupp, Director of the Griffith Observatory in Los Angeles, and Richard Walker, an astronomer at the U.S. Naval Observatory in Flagstaff, Arizona.[208] The projects involve a lot of cross-disciplinary research to obtain precise surveys of geological data inside the crater, the topological measures of the site, insight into the drainage systems of the crater, the seasonal biological processes and metereological conditions of the area, astrophysical data concerning the position of the sun or moon at such and such a day, and so forth. The growing concerns on the part of environmentalists not to disturb the delicate balance of the desert have also been taken into account. Yet the purpose of the scientific data gathered is completely geared toward research into the nature of perception of sky phenomena. By making perceptual experiences the sole 'object' of the work, the distinctive relationship between viewer (subject) and object (a non-objective visual phenomenon, i.e. reality) has changed as well. James Turrell has created an experience as real 'object,' that can be interpreted subjectively as well. One can say that this 'reality' exists only in the eyes of the viewer-participant. Perceptual experiences are known to be only relatively objective, i.e. what a person sees depends partially on what he or she knows, and how a thing, a problem, or reality, is approached. Whether intentional or not, it raises the question of the objectivity of scientific experience. I should like to recall Werner Heisenberg's discovery of the Uncertainty Principle which maintains that the presence of the scientist always affects the result of a scientific experiment.

To summarize, the *Roden Crater* is a completely thought-out interdisciplinary research project, constructed to create a systematically composed layering of events in time and space, to be experienced by the visitor as environmental, physical as well as perceptual phenomena. This does not mean, however, that this experiential 'program' is necessarily chronological, or linear. On the contrary, the layering and combination of 'events' or experiences should be interpreted as an interrelated series of data, or givens, from which the visitor may select and create his 'compositions.'

We should be careful to ascribe religious or esoteric aspirations to these works, even if some of Turrell's pronouncements make it tempting to do so. Artists like Turrell and Holt, whose interdisciplinary research and mode of working brings them in touch with current developments in astrophysics, such as research into galactic phenomena, black holes, and red and blue shift phenomena, share is a common concern for the cosmic qualities of the earth at large, even if the works themselves do not share any stylistic or iconographic similarities.[209] James Turrell wants to stimulate mankind's awareness of the position of the earth in the universe by making certain phenomena in the sky visible, to which one would not normally pay attention. Nancy Holt does so too, as we will see. Their concerns have reached a high level of complexity as regards the formal, perceptual, symbolic, environmental and scientific aspects, which one does not find in the earthworks of the sixties. Even if these concerns may be fed by a renewed interest in the large 'earthworks' of past cultures, they are also firmly based on current state-of-the-art research, although this search appears to be principally guided by the 'holistic' world view, wherein everything is related in some way to everything else, and wherein the cosmos functions as a unified system. During the 1980s, there was a genuine interest in these ideas, parallel to the newly published books of respectable scientists like Fritjof Capra, Gary Zukaw, Rupert Sheldrake, David Bohm, who all connected recent concepts in the physical sciences with thoughts in eastern philosophy.

NANCY HOLT: Contextual Site Works

Astrophysical Connections

Nancy Holt began her artistic career as a photographer, moved to film and video in the early seventies and changed again to work in the natural and urban environments in 1972, which has remained her subject. She considers museum and gallery conditions unsuitable for her work.[210] The artist's first major permanent work, *Sun Tunnels,* in the Utah Desert (1973-1976) still has the qualities of a sculptural earthwork: a large scale, object-like construction, set in a desert landscape away from the cultivated environment, difficult to reach. Since then, she has situated her work in cultivated and urban environments, and the relationship between the site (whether chosen or commissioned) and the constructions has grown stronger, and more complex. The projects now incorporate a layering of elements that connect them with the environment at large: the site's cultural and natural history, its social utilization, and ecological concerns. The continuous thread is her interest in visualizing the astrophysical correlations between the location of the work and the solar system.

The *Stone Enclosure: Rock Rings* (Western Washington University Grounds, Bellingham, WA, 1977-1978) consists of an inner and an outer ring of stone from the area (brown mountain stone). There are formal qualities in the design of the rings, such as the spatiality of the holes whose positions are related to those of certain stars. Holt has extensively described the work herself, because she considers it an aspect of her work to provide the information necessary to understand it more fully. Holt: "The four arches in the work are aligned on the North-South axis as calculated from the North Star, Polaris, in a way similar to the way the Northwest Coast navigators plot the courses of their ships. Being aligned with the North Star, the work relates to a point in the universe, a dead center, a true North. The work becomes astrally fixed on Earth. The 12 poles are aligned NE-SW, E-W, NW-SE, spatially orienting the landscape and bringing the horizon back to the center of the work. Through the holes and arches various aspects of the landscape are framed and come into focus - part of a mountain, the trunk and branches of a fir tree, a slice of road, an asphalt walkway, dense woods, a small knoll. The views are chosen - using a circular frame mounted at eye level and placed on various directional axes, I could see exactly what would be viewed. ... Rock Rings is meant to be seen from many points of view, from the inside and the outside. The work is constantly changing perceptually as I move from place to place."[211] The spectator has to move in and about the piece, has to spend time with it to see the changes of the light during the day, or locate the position of the stars at night. Nancy Holt consulted an astrophysicist and a geologist on the piece, who did the calculations. Also, she worked for three months with an experienced stonemason and craftsman, Al Poynter, on the construction. Similarly, in *Star-Crossed* (Miami University Art Museum, Oxford, Ohio, 1979-1981) she set out to create relationships between visible and, under normal conditions, invisible aspects of the site, exploring its astrophysical relations. It centers around the fact that Oxford is one of the places in the world where the magnetic north and astronomical north fall together. The artist designed two tunnels, a small tunnel positioned along the north-south axis, and a big one positioned east-west, crossing each other on an elevated ramp. At the end of the small tunnel is an oval pool. However, when the visitor walks up the ramp and looks through the small tunnel, that oval pool appears a circle. "It fits perfectly the field of vision framed by the circular tunnel... When looking into the pool, one is also looking down to the ground, and yet seeing tops of the trees and the sky and the stars - so the sculpture inverts the world," says Holt.[212] Making visible one reality, the sky, through reflection or mirroring, she connects the two, as it were, perceptually. Although her approach is different from James Turrell's, there are similarities in the way their works intend to make the viewer perceive celestial phenomena.

Design of an Urban Landscape Park

During the past decade her attention has shifted to include the whole spectrum of outdoor issues reflecting the situation of the site. It has become the pretext for the creation of a work that intends to make the viewer aware of the existence of a larger context in which he or she partakes. Over the years the projects grew larger and larger and simultaneously the layers of meaning became more and more intricate.

For *Dark Star Park* (Rosslyn, Virginia, 1979-1984) Nancy Holt was invited to create a 'sculpture' in a park, and ended up designing the whole park, taking on the role of a landscape architect. The site encompassed a parking lot, the plaza of an office building, an island three-lane traffic intersection and a small park, all looking desolate. The designated site was already part of an urban renewal project. As always during preparatory stages, Holt collaborated with the urban planners and sought the assistance of an astrophysicist. She also studied the local history and found that on August 1 in 1860 a certain William Henry Ross acquired the land that became Rosslyn. The elements which she chose are comprised of five spheric balls, sets of poles and reflecting pools, tunnels and curved walkways. Considering the pedestrian and automobile traffic, the pathways, stairs and tunnels are both functional, and socially interactive. Yet again, the work is related to the site in a specific way. The shadows cast by the spheres and the poles will line up with the shadow patterns on the ground at 9.32 a.m. each year on August 1, "to merge historical time with the cyclical time of the sun." The different elements are aligned to cause perceptual phenomena or illusions, for example the spheres may seem to eclipse depending on the visitor's route. Here also, the themes that dominate this work are history and time, and the sky and stars. The viewer perceives the work and the environmental situation as a complex system of literal and mental pathways along which information about the natural aspects of the site as well as about the internal structure of the piece 'travels,' as it were.[213]

Sky Mound

Nancy Holt's most encompassing work to date is *Sky Mound*, a 57-acre landfill in the meadowlands of Hackensack, New Jersey (1985-1994). It is a $ 600,000 project, paid in part by the Hackensack Meadowlands Development Commission, which also commissioned the piece.[214] *Sky Mound* can functionally be designated as land reclamation, in order to "transform a damaged site into a life-generating place."[215] Formally it refers to ancient observatories, whose central axes visualize annually recurring cycles, such as solar and lunar solstices. In this way she directs attention toward their original function as centers of astrophysical research. As "calendars" they also dictated the annual rhythms of life, when seeds were to be sown or crops had to be harvested. The mounds, with staircases and platforms to observe the stars, remind us of the American-Indian tradition. From a central point, paths radiate like rays of the sun. The alignment of the paths, poles and mounds enable the observer to view the changing patterns of the nightly stars and other celestial phenomena.

Where Herbert Bayer's *Mill Creek Canyon Earthworks* (Seattle, WA, 1979-1982) follow the course of the stream and the exisiting contours of the surrounding hills and meadows to create an environment that is conducive to the public's recreational activities, Nancy Holt adds an additional layer of symbolic references by aligning pathways, mounds and poles to make them form a structural overlay representing almost a mirror image of and means to visualize celestial phenomena. The lay-out connects the earthly elements with 'heavenly' or 'cosmic' ones. The ecological aspects are taken care of by the creation of a pond draining the surface water. It might also be a refuge for birds and wildlife. As a remembrance of the landfill's previous function, a methane flame will burn on holiday evenings as as symbol for *Sky Mound*'s recovering and recycling function. As a matter of fact, *Sky Mound* is the result of collaborative efforts. Holt worked in this project with landscape architects Cassandra Wilday and Katherine Weidel and archeoastronomer James Mavor, among others. Nancy Holt's work is of an interdisciplinary nature. She works with astronomers, geologists, and biologists to obtain the necessary information. She collaborates with craftsmen and technicians on the construction. Because her projects belong to the public context, working outdoors in this manner has changed her role as an artist/sculptor. The artist's work can be defined as a more functional type of art activity, which reaches far beyond the placement of an object in the park. Not many artists are capable of handling the complexities involved in such projects, such as dealing with administrators, local officials, politicians, sponsors and funding agencies. Apart from *Sun Tunnels,* which was funded by the artist herself, most of the artist's works have been commissioned. In Holt's case, these have been mostly university and college art museums, and occasionally federal, state or city arts commissions.

Chapter 7

Robert Irwin - Art as Inquiry

Transgressing the Frame

Robert Irwin made it his first objective to develop new visions about what the visual arts should be concerned with.[216] Since the seventies he has set out to put these thoughts into his art, writing and lectures. Central to Irwin's activities became the research into perceptual processes, and the act of seeing or perceiving itself became the subject of his work. The road that he followed was a step-by-step investigation to rid himself of the picture frame. His first steps began with experiments in different color combinations and hues. By slightly curving the surface of the canvas he attempted to dissolve the edges of paintings, so that it was barely visible where they ended and where the wall began. These became circular, convex aluminum discs, sprayed with a matte acrylic paint and mounted on a concealed male-female tubular arm, twenty inches forward and parallel to the wall. This was in 1966 and 1967. The dimensions of the disc and the space were exactly measured. The discs were cross-lit from four corners by incandescent lamps of equal intensity. The result of this cross-illumination combined with the ambient light was that the shadow, the disc and the outer area of the illuminated wall were seen as an entity. In this kind of indeterminate field, disc and light became almost interchangeable in terms of physicality. The perceptual ambiguities forced the viewer into a process of determining what he or she was actually seeing, or on another level, to consider the nature of perception itself. The next series of transparent discs, says Irwin, "made it easy to make that transition between physical and so-called non-physical." What you had there was an alteration in the whole promise of perception, questioning your concepts of 'real.' In fact, these dematerialized objects questioned the given cultural framework of our perceptual habits, showing the relativity and transiency of what appears solid matter.[217]

Robert Irwin was born in Long Beach, California in 1928, and resides in San Diego.[218] He began his artistic career as an abstract expressionist painter, and exhibited his paintings since 1952. Like many artists at the time, Irwin became dissatisfied with the canvas, and questioned the parameters of the art object. During this early part of his career Irwin moved among a group of artists associated with the Ferus Gallery.[219] Because of his subsequent investigations into the figure-ground relationships which resulted in the discs, critics grouped him with a number of artists whose concerns appeared similar, and who became known as 'light and space' artists. Among them were Michael Asher, Maria Nordman, Eric Orr, Michael Brewster and also James Turrell. One of their principal aims was then to eliminate the art object from the gallery, using only either natural or articificial light sources, or, as in Brewster's case, sound, to affect the perceptual qualities of the space. Initially Irwin also followed this path.

At the beginning of this direction stood the artist's involvement with perceptual and experimental psychologists during the *Art and Technology Program*, mentioned above. These experiments widened his interests into an investigation of the human responses to environmental conditions in general. As a consequence, he gave up his studio in Venice in 1970 and spent most of the next two years traveling and lecturing to develop his thoughts on art and circumscribe his ideas in words and writing. Finally, he stopped making objects altogether. Irwin now reduced his material to a (nylon) scrim and/or (natural) light with which he created visual experiences that involved a process of visual illusions depending on the position of the viewer or the situations that changed continuously in respect of the light that entered the space from the outside. These *Scrim Pieces* were exhibited for the first time in 1975. Critic Roberta Smith described her experience viewing one of them, *Scrim V*, at the Museum of Contemporary Art in Chicago: "It is essentially a closed-off, wedge shaped corridor, whose sides start at two pillars near the museum's entrance and meet at a far wall, some 80 feet away. It presents numerous views; you could look both into and through it to the large empty area of the gallery space that it blocked off. After the first, symmetrical encounter at the wide end, nothing was constant when you walked along the side. The two panels of scrim were always at a different distance from each other and from the wall that they generally paralleled. As you moved, whole chunks of architecture and space were constantly dropping in and out of view. At times you looked right through the volume but still couldn't really locate the far wall; at other times the scrim itself seemed opaque and solid."[220] The central theme of the work was the perceptual experience of the visitor, whose experience would continuously change during the day and the time spent with the piece. For as the eyes adjusted to the space, the viewer would perceive more and differently. Studies in perception have shown empirically that human eyes take at least ten minutes to adjust to new light conditions. The 'reality' of seeing thus became a complex set of space and time relationships.[221]

A Site-determined Approach

Since 1977, Irwin has been making site-determined works, and has not often been shown in museums and galleries.[222] In these so-called site works, the artist has continuously developed and redefined this approach as a method of enquiry, as it were, and he simultaneously lectured and wrote his theory of "art in public places." In *Being and Circumstance*, published in1985 in conjunction with his exhibition at the San Francisco Museum of Modern Art and The Pace Gallery in New York, he explains this vision. Robert Irwin defines four types of site-related 'sculptures,' as distinguished from the generic term public art: 1. site-dominant, 2. site-adjusted, 3. site-specific and 4. site-conditioned or site-determined works.[223]

His categorization has of course been criticized, if only because his distinctions often overlap. Yet the definitions are often used and quoted now, because they do indeed help to qualify different levels of relationships between the work of art and the surrounding space. For Irwin there is a large distinction between the work growing as it were out of the site, and the work that is conceived in the studio, even if considerations of place and placement, of relationships of material, size and scale to the surrounding space come into play. He describes this category as a procedure which breaks with "the conventions of abstract referencing of content, historical lineage, oeuvre of the artist, style, etc., implicit in the other three categories," and crosses "the conventional boundaries of art vis-à-vis architecture, landscape, city planning, utility, and so forth, reducing such quantitative recognitions (measures and categories) to secondary importance." By placing the individual observer in context, at the crux of the determining process, it follows "the principles of phenomenal, conditional, and responsive art."[224] What makes Irwin's book particularly important in other respects, is that the artist has described the whole process in which a sculptor becomes involved when he decides to create this type of work. He enumerates his experiences and problems encountered and makes suggestions to avoid risks, and mistakes that happen out of ignorance with common affairs surrounding the construction of public works.

Inquiry into the Hidden Structures of Art

During the course of their career, almost all the 'light and space' colleagues began to treat the specifics of the location where the art work was to be as an elementary part of it. Yet Robert Irwin stands apart in that his work truly comes out of the characteristics of the space chosen or commissioned, in claiming that his starting point is like a "tabula rasa." Always he sets out with a question about the unknown, which he calls "the art of pure inquiry," guided by "the desire to know."[225] His inquiries into the making of art itself became a committed investigation into the system of art, and opened the way for a redefinition of art itself.

It is exactly at this intersection where his road deviates from the directions taken by 'light and space' sculptors. When James Turrell very precisely creates the conditions for perceptual and physical experiences of lunar or stellar phenomena occurring at specific intervals of the solar system, for example, he still creates an artificial situation - even if it heightens the awareness of the visitor, and even if he exerts the utmost care not to disturb the ecology of the environment. Turrell already had conceived his idea of an observatory beforehand, although he developed the total concept during the working process. Irwin always approaches the site without preconceived notions, and he does not impose his art on the site. Both Turrell and Christo sell their preparatory drawings and plans to finance their projects. For Christo, the art work starts with the conception of the idea. The whole process of preparations, of selling the concept, the fundraising and publicity efforts, the production phases and finally its removal, are all part of the work. Irwin, on the contrary, perceives these aspects as being external to the work.

Richard Serra's monuments (such as *Tilted Arc*, now removed) may be designed for their setting, but they are in fact large objects trying to dominate and compete with the surroundings. For Irwin, they still belong the classical tradition of sculptural objects. Serra's *Tilted Arc* (1981) has been compared to Irwin's *Tilted Planes* (1978, unrealized), presenting two examples of a different approach in public art. *Tilted Arc* made for Javits Plaza, downtown Manhattan, dominated the site with its large statuary of slightly bent corten steel plate units. One could not avoid its presence, had to walk around it, as the sculpture split the plaza in two, almost. The office workers felt that the art had invaded their space and protested, with the result known. In contrast, *Tilted Planes*, a project for the Oval Mall at Ohio State University, Columbus, would have been almost invisible. Using the existing pathways and traverses across the mall, Irwin proposed to slightly tilt a number of the planes, so that the Oval became, as it were, three-dimensional, accentuating the paths and intersections. Irwin: "To me, it was already a piece of sculpture. It had all the dimensions and all the properties of a piece of sculpture: physical divisions, both organic and geometric, participation of people, the kinetics of movement. It was already operative in that way."[226] Starting from Irwin's premises, Walter De Maria's *Lightning Field* in New Mexico (1974-1977) would be designated somewhere between a site-adjusted and a site-specific work. The artist placed four hundred stainless steel poles, 18 feet high, in a rectangular grid, measuring 1 mile by 1 kilometer, in an area with frequent lightning and storms at certain periods of the year, so that the poles act as lightning rods. At other times the rods are barely visible. Although enhancing specific, temporary phenomena in the natural environment, nature seems only partially integral to the work, though. Richard Fleischner, an artist who works in the field of public sculpture, might be more in line with Irwin's fourth category. Fleischner himself has described the development of his proposal for the Jerome Wiesner Building, familiarly called Media Laboratory, at the Massachusetts Institute of Technology, Cambridge, as a sculpture that evolved completely out of the complexities of the site, by spending hours walking around, sitting and watching. He was invited to design an inner courtyard section for the newly built Media Laboratory, to be designed by the architecture firm of I.M. Pei & Partners. In the catalogue interview, he mentions that his project soon expanded to comprise just about every inch of the surrounding space, and all the development aspects necessary for its execution. He decided which areas were to be paved and which planted with grasses and plants. He even ended up designing the lighting and seating, normally the realm of the landscape architect. He was in close contact with the architect's office working with all the parties involved, from Wiesner, the architect and the contractors, to the tradespeople and laborers. Fleischner perceives his task as a problem-solving situation, whereby he goes about it as an inquiry, looking for different solutions possible, ordering and structuring until the desired result has been achieved, and his approach resembles Irwin's in that respect.[227]

Robert Irwin's starting point requires him to leave all the old notions behind, in order to 'surrender' to the selected environment at large. Clearly, Irwin's research can only result in an 'object' contextualized in the site. It should not surprise us that his works do not have much stylistic or formal congruences. Their only common quality is their starting point - the connection to the site - which often results in a virtual disappearance into the site. The *Filigreed Line* made for Wellesley College, Massachusetts (1979), consists of a stainless steel line, running along a ridge of grass near a lake, in which a pattern of leaflike forms is cut. The height of the line varies between 2 feet and 2 inches from the surface of the ground. The 'line' keeps changing visually with the light of the day, and the seasons of the year. Because the line oscillates before the viewer's eye, who needs to focus and re-focus, he or she is made aware of the nature of perception itself. The *Two Running Violet V Forms* at the University of San Diego, California, has a similar quality. The two crossing blue-violet, plastic coated wire fences fixed with high poles and sited in between the eucalyptus groves hover between appearance and disappearance, continually changing with the light, the seasons and the position of the spectator. The emphasis is on the integration of the material in the landscape and the perceptual process as experience of the whole, to direct the viewer's attention to the whole environment away from the art 'object.'

Although never realized, Irwin's proposal for the expansion of Miami International Airport is undoubtedly his most ambitious project to date. The development of the plan is extensively described in the exhibition catalogue published by the Museum of Contemporary Art in Los Angeles in 1993. It is also the first full-scale survey of Robert Irwin's career. I shall therefore give a concise narration of its history, with quotations from Lawrence Weschler's interview with Irwin, as they are telling about the artist's approach toward his subject. But mainly because it seems to stand at a crossroads in a continuing development away from the art object and its well-defined boundaries, away from the 'figure' as focal point towards an inclusion of the background, and subsequently the surrounding space, whose expansion (temporarily?) ends with the engagement of the whole environment, or one could even say the world at large.

Being invited by the Miami city arts commission, headed by Patricia Fuller at the time, Irwin flew to Miami to look into the situation. "The airport was chaotic, incredibly chaotic. One thing it sure as hell didn't need was more clutter - another object: art. What it was crying out for, though, was an overall approach. And in fact it seemed a perfect occasion for looking at the situation of public art generally, which is in similar disarray. An airport was a perfect late-twentieth-century case study ... teeming with transience and cross-purposes and disarray."[228] Robert Irwin took with him Ed Wortz, a long-time friend with whom he had worked before, and Coy Howard. They spent a week in the airport hotel to explore the place. His conclusion was that the airport buildings, largely built in the fifties, were of a totally nondescript character, without any sense of location, thus acutally in need of a complete overhaul. Irwin, Howard, and Wortz spent hours, days, looking into the characteristics and possible opportunities of the site and came up with a flow chart, dividing the airport in different zones the passengers passed through. They observed that the zones consisted of six phases which repeated themselves consistently: arriving, passage, seeking, finding, waiting, and leaving. Also, he felt that these zones, i.e. parking-lots, check-in counters and waiting areas, corridors and passageways, baggage claim area etc. should be blended more. Thus Irwin started to work and eventually presented the arts commission and the airport's board with a three-part proposal after several months of work, which was to become his *Arts Enrichment Master Plan* and was to involve years of work (1986-1990).

The basic element of this master plan was the presentation of a completely new concept of traffic flow from the arrival of passengers and visitors, via check-in to take-off and vice versa. Its main theme was the enhancement of the quality of the specific sites, creating as he calls it a sense of location, so that people would know where they are, move quietly from one zone to the next and leave the usual disquieting haste and nervousness behind. In Irwin's vision this meant "developing an aesthetics or quality" of site. In practice it boiled down to proposing a thorough rehabilitation of all the passageways and locations of the airport. He never intended to re-design the whole airport himself, but thought of inviting the appropriate artists for the different sites. Only the entrance section he reserved for himself. The essence of the proposal could not be reflected better than in Weschler's description: "He was proposing to tear out the fairly dilapidated two-story parking garage at the center of the entire facility, surrounded by entries to all the various terminals, and to replace it with a lush cypress grove. The grove would be layered, with fabulously varied native fern and birds rising from the marshy lake below - meanwhile, arching through the canopy, pedestrian skyways would link the terminals to each other and to the relocated parking structure beyond. There would be benches and cafés, zones for quiet contemplation during those long layovers." Furthermore, "his project assumes that the airport, the first and last part of the city a traveller experiences, should in some way emblematize the city, rather than serve some impersonal outskirt function, architecturally everywhere and nowhere."[229] Irwin's vision was to have the passenger on his way to and from the airport pass a kind of Central Park landscape reflecting the abundance of Florida's flora and fauna, thereby 'planting' the first impression in the mind of the incoming visitor, and 'leaving' a lasting experience and memory for the departing passenger.

Initially meeting - of course - with the usual suspicion, Irwin had to put forth all his techniques and powers of persuasion to convince the city and airport administration. He had to surmount numerous bureaucratic hurdles and opposition from, for example, the liability department. However, Dick Judy, airport director at the time, became more and more interested, and by 1990 he seemed almost convinced and about to give his go-ahead for the whole project. At that crucial point in time, Judy got fired, or rather had to 'resign.' With the most important person whose trust and respect he had finally gained after three years gone, the project soon died.

Although the artist generally does not purposively set out to restore an older flora typical of a certain climate, or a once existing bioregional diversity, he is quite sensitive to the ecological characteristics of the location. For the proposed Miami airport entrance corridor, in Florida's Dade county, cypresses, palmettos, and reeds, taking into account the local waterscape situation, would have been planted. For *Sentinel Plaza* (Sentinel Plaza, Pasadena, California, 1990) Irwin chose small desert plants and cacti, echevaria, to transfer the message that water is precious in this climate. So does the double-trunked sycamore, which appears to have been there for ages. The Indians used the trunk of this tree to build their kavaks. At the center of the piece stands a granite column ornamented with the petals of lotus (?) flowers and a blue (police?) lamp on top, reminiscent of the federal eagle. At its base extends a thin trough-like basin ending in a small circular 'fountain.' Behind the plaza is indeed a police station. *Sentinel Plaza* is an example of how Irwin connects the separate existing elements and their respective symbolic connotations into a layered 'system' of meaning: the surrounding buildings, architectural style and function included, the earth and climatic conditions determining the choice of plants and trees, and finally, on a more abstract level, the inclusion of time or age.

Robert Irwin has called his approach one of response to the conditions of the selected or commissioned site. He always starts from the premise that a work can only come into being by asking questions - about the past and future of the site, its social, environmental, political, cultural ramifications. It is a constant inquiry, but no longer about the work's function as art but as life, as a living environment. His plans are designed with the intent to enhance the spatial potential, including the aesthetic quality of the location, thereby - hopefully - creating a "sense of awareness." Irwin: "The art of pure inquiry is an open interface between the pure subject - all that is out there - and the pure potential of the individual perceiver - all that is in here. Where the strength (clarity) of this inquiry lies is in its single motive - the desire to know."[230]

The recognition of the hand of the artist has become unimportant to Irwin. His major concern is still to make us "see again," thereby continuing his research into perceptual experience which he had started in the sixties, but now expanded to include the context of a cultural, social and political environment. Irwin: "Basically it's just to make you a little more aware than you were the day before of how beautiful the world is. It's not saying that I know what the world should look like. It's not that I'm rebuilding the world. Basically what artists do is to teach you how to exercise your own potential - they always have, that's the one thread that goes all the way through."[231]

Chapter 8

The Artist as Ecologist

ALAN SONFIST: Nature as Material

Since 1965, Alan Sonfist has collected threatened species to bring them safely to another place, called *Element Selections.* Although he graduated in painting and art education at Western Illinois University in 1967, he already stopped painting upon his return to New York in 1968 and started growing micro-organisms. The themes that ran through the early process works were intended to make the spectator aware of 1. nature's recurring time-cycles, patterns and structures. 2. the natural and man-made environmental changes that have taken place throughout the history of the earth. His experiments with micro-organisms resulted in a series of canvases painted with water. Soon, mildew patterns appeared on the stretched linnen of the quickly ageing canvases, which literally deteriorated before one's eyes in a short period of time.

For another temporary piece he brought snails into the gallery (Institute of Contemporary Art, London, 1972). The *Snail Enclosure* showed two dozen black snails moving about in a square containment. The snails left trails of gleaming mucus behind them that over a period of time formed an increasingly tangled network of intersecting lines. In these works Sonfist as it were (re)staged natural conditions in a gallery situation. He had already used a similar approach in the *Colony of Army Ants* (1972). In *An Autobiography of Alan Sonfist,* the artist explained: "For three weeks I searched for a colony of army ants in a jungle in Central America. They were housed in a rectangular enclosure to present the functioning of an entire living society, rendering it visible. During the exhibition, I placed food in various predetermined geometric patterns. The ants varied these patterns according to their own pragmatic specific paths. Each of my predetermined feeding patterns was seen in a drawing on the wall. The "drawings" directly beneath them displayed the patterns that the ants had created while traveling in search of food. At the end of the show, in response to the totality of the artist-imposed patterns, the ants will have selected and stabilized their own predominant patterns."[232] Although Sonfist never expressed his antagonism toward the art world in writing, his objectives would gradually move his activities out of the museum and gallery situation, so that these works can be interpreted as statements against the notion of paintings as precious art objects. Nature and natural phenomena became Alan Sonfist's media, including himself, and the earth with its history of natural processes, cycles, as well as the changes caused by human intervention.

Consequently, he was invited to participate in exhibitions dealing with natural elements like the earth, air or processes, and with the landscape at large.[233] The artist's use of his body as part of nature occasionally placed him in the Body Art movement, or in something like 'ritual in art.' In these works; some critics saw references to Duchamp (found objects) or Minimal Art (comparable formal aspects in, for example, the crystal works). His two fellowships at the Center for Advanced Visual Studies at MIT in 1971-1973 and 1976-1977 made sure that he also participated in exhibitions which connected environmental with technological issues. His early preoccupation with natural systems led to a comparison with Hans Haacke, as a matter of course. Sonfist has never allied himself with the systems approach in the ways it was used by Haacke and defined by Burnham, although the works described above could be interpreted as such. Sonfist has always shied away from any categorization.[234] It is a matter of fact, though; that the terminology used by Sonfist himself to exemplify his procedures is derived from systems theory, and that he was familiar with its concepts. Writing about the issues that concerned him became a way to explain the larger and interdisciplinary context of the work beyond the frame of purely art-historical references.

Crystal Globe (c. 1966) was a large glass sphere containing natural mineral crystals. As environmental conditions changed under the influence of heat and light, crystals vaporized into a purplish gas which then re-crystallized. Although the process itself was predetermined, the kind of configurations kept changing, making up the random or chance element. *Crystalline Enclosure* (1969) was a lucite globe form again containing crystals that changed form and place continually in response to the temperature and air currents in the surrounding atmosphere. In a stable environment there is little or no change. A disruption of the environment (just the presence of spectators would cause changes in air currents, for example) would cause the system to move from a state a relative equilibrium into disequilibrium. Being a self-regulating system, the system would react on a disruption of the environment by trying to move back again into a (i.e. another) state of equilibrium through feedback or self-regulation. Sonfist explained: "The growth of the work is characterized by patterns of indeterminacy, since a multiplicity of natural variables creates unforeseen occurrences."[235] His experiments with crystal minerals a few years earlier had made visible certain crystalline patterns of change that recur continually in nature, but of which we are seldom aware. "I want to make visible the invisible phenomena of the universe," explained Sonfist in an interview.[236]

When Alan Sonfist experimented with natural phenomena, whether they were growth processes of crystals and crystalline structures, or chaotic patterns caused by micro-organisms, or relationships within a closed living system of army ants, there were always a number of elements involved that acted upon one another, constituting the overall framework within which the pieces had to be perceived. The most important elements were the visualization of: 1. a normally invisible micro-world - 2. patterns and structures, whether chaotic and random, or organized in an orderly manner - 3. processes in time, whether present, past, or moving toward the future. There are conceptual similarities in the works of Haacke, Smithson and Sonfist made around that time, but rather than comparing their works in terms of differences and congruences, I prefer to look at them as timely expressions of ideas and attitudes prevailing at the time. Both Smithson and Sonfist explored the nature of crystals, for example. Thematic congruency in Haacke's, Smithson's and Sonfist's work with such concerns as time/space, change/process, randomness, making visible the invisible, references to the geological, biological, and human history of the site in terms of interdependent entities may be based more on similar interests in current scientific developments and ideas than on mutual influences. In a much later interview with Michael Auping, Sonfist responded that he was not aware of Smithson's work at the time. "It's difficult to speak of influence because I was very involved with my work at the same time they were making theirs. ... The way they work is so totally different from the type of work I'm involved in. They are imposing themselves on an area that has not been touched by humans; and I'm more concerned about *not* going into areas like that."[237] What he actually says here is that he feels that his attitude toward nature differed completely from that of the early earthworks. Sonfist and Smithson apparently moved in different art circles, so it may very well be that they were not very aware of each other's activities in the late sixties. However, there were a few similar starting points that served as working methods to give a new meaning to art in the work of both artists, while their objectives differed considerably. Both artists set out to search for a different position as artists by moving outdoors, and professed a functional type of art away from the museum/gallery situation. It was an era of new technological developments, and the new ideas in the sciences floated around in many different disciplines. This interest would unavoidably bring them in touch with the thoughts that have been discussed in previous chapters. It was also a period of renewed interest in the natural environment, which developed into an ecology movement.

The works which Sonfist made during the seventies reveal internal concerns with interdependent patterns and structures that are created as process, over a period of time. They were the result of staged artificial situations, surrogates of the natural condition, resembling laboratory conditions. If time as a literal process constituted a major element in his early work, the awareness of history, that is memory of things past, grew in importance over the years. His visualization of time now no longer followed the Newtonian definition of time: it could be both linear (the sequential time of a process) and cyclical (recurring cycles in nature); it might condense or stretch. Human beings participate in an earthly 'universal' process, of which one only experiences a fragment. By recalling or recreating environmental situations as they happened in the past, the artist wanted to point at "the isolatedness of one visual perception in a moment of time. So that any one visual image must be seen as only one segment in a total process. Any one form must be seen as an essence in a continuum that is shaped by other, perhaps visually absent, factors. Landscapes present the fragmentation of a continuum."[238]

Sonfist's method is one of inquiry and problem solving, to be compared with scientific methods. The work is the result of structured observations and inquiry. Robert Horvitz, virtually the only writer who has perceived this, wrote: "The problems he has increasingly focused on are akin to those encountered by the painter of landscapes: how to select an intelligible order from a wealth of physical appearances which cannot be fully grasped by either himself or those to whom he wishes to communicate. Decisions must be made on the basis of incomplete information about the meaningfulness of certain observations at the expense of others. These problems are also encountered in scientific research."[239] Horvitz' "scientific" interpretation describes the artist's explorations as "a broad cross section of substances, many of them dynamic and most of them involved with the translation of energy from one form into another. ... Sonfist especially wishes to bring out those patterns of behavior which recur endlessly in nature: the transmutability of matter and energy, the hierarchical layering of forces, feedback and self-regulation, multiphase equilibrium, etc."[240] Interestingly, his terminology, too, is derived from systems analysis and cybernetics.

Human beings have used and transformed nature into landscape from the start. We are the only ones who did so consciously, and we are also the only ones on this planet who can decide how to proceed. If the future of the planet ultimately remains uncertain, it is mankind who can and will determine the earth's future. Sonfist: "Most of my pieces deal with this idea in an ecological sense in which I present the change in form caused to bring change in another related form."[241]

Nature as a Public Monument

This is the point of view from which Sonfist developed his 'theory' of natural phenomena as the true public monuments: "Public monuments traditionally have celebrated events in human history - acts or humans of importance to the whole community. Now, as we perceive our dependence on nature, the concept of community expands to include non-human elements, and civic monuments should honor and celebrate the life and acts of another part of the community: natural phenomena."[242] The whole complex of a site's natural history, geology, biology, ecology, human interventions, etc., became the subject of Sonfist's outdoor public monuments. Its foremost feature is that there always exists a direct correlation between the site and the work of art to the extent that, says Sonfist, "each particular context suggests the form for my sculpture."[243] *Rock Monument,* commissioned by the Albright-Knox Art Gallery, Buffalo (1978), is an installation of rocks that were all selected from the Buffalo area, and positioned in the same relationship as they were originally found. For Sonfist the geological strata, which is geological time, tell something about the history of the area. *Rock Monument* condenses a time span of billions of years in the present, brings together data of a larger area on a small site. There is also a relationship to the museum with its collection of different periods of art, representing a segment of the history of time, so that the piece visualizes patterns that would normally have remained invisible. How different Alan Sonfist's approach is from that of other sculptors may become clearer when one compares the Buffalo *Rock Monument* with a superficially similar work by Carl Andre, *Stone Field Sculpture* (Hartford, CT, 1977). In *Stone Field Sculpture* boulders have been lined up triangularly, the sizes increasing with the diminishing length of the rows. Andre's sculpture is about place and numerics. The work's only relationship to the site is that it is located there, whereas Sonfist's work comes out of the site's multiply layered context. Michael Heizer's *Adjacent, Against, Upon* (Seattle, WA, 1976), falls into the same category. Boulders have been placed and stacked, cut or left uncut to indicate the proportion, size or spatiality. Heizer's sculpture still has the monumentality of his architectural complexes in the desert landscape of Nevada. Only recently has he completed a work that can be considered as land reclamation, called *Effigy Tumuli* (1990), a series of earth mounds with configurations of animals referring to American-Indian mythology.[244]

When Sonfist needs the assistance of professionals in other disciplines, or technological help, he will approach the person that seems most suitable to perform the task. For *Colony of Army Ants* he contacted the leading authority on army ants, Howard Topoff. He worked on *Rock Monument* with geologist Richard Laub, of the Buffalo Museum of Science, who accompanied the artist on his rock search. For the ecological *Time Landscape* he had the site analyzed by a team of scientists, including a biologist, a botanist, a chemist, a geologist and a city planner, to determine what natural phenomena existed previously and what effects the present atmospheric pollution and noise would exert on such an original environment when reconstituted. He even set up a nonprofit organization - Conditions, Inc. - to legalize his position.[245]

Time Landscapes

Time Landscape in New York City (1965-1978) seems to bring together the major concerns that run through Sonfist's oeuvre to date. It is the recreation of a precolonial forest of 200x40 ft. on a formerly empty lot on La Guardia Place between Houston and Bleecker Streets.

Although the actual implementation of the work began in 1977, Sonfist had already approached the Board of the Metropolitan Museum of Art to consider a segment of *Time Landscape* on a part of the museum grounds in 1970. In reality, the project already began with the collection of the *Element Selections*, a series continuing until today. The growth project shows the simultaneous time of three stages of a precolonial forest similar to the one that existed on the site prior to the colonization of the land. Changing and growing according to its own cycles, it consists of trees, shrubs, grasses, and wildflowers that used to grow in the area but became extinct when civilization spread, causing its almost complete deforestation, and is now a living part of the community. *Time Landscape* brings together past and present, referring to the times when New York used to be forest. The recreation of the precolonial situation is not just a romantic gesture, made from nostalgia for past times, but refers to current problematic environmental conditions. In addition, it confronts the passers-by with the process of time, as a continuum of which we only perceive a moment. Finally, it is symbolic for the idea of memory (history) and life(-cycles). Sonfist: "The history of a place includes the history of its natural environment. Within an area of landmark houses, it will be shown that it is possible to create landmark nature. *Time Landscape* renews the city's natural environment just as architects renew its architecture. The concept of evolutionary change can be applied to any site of the city. This is a pilot project for this kind of reconstruction and documentation that can coincide with new building in the city."[246] He finished the planting of his most renowned outdoor work to date in 1978.

One may compare *Time Landscape* with a major urban design plan. Collecting the necessary scientific data is only part of the effort. In order to execute such plans, one has to weave one's way through the city authorities, deal with politicians, community groups and real estate interests, hire lawyers to establish one's legal position, etc. Organizing and supervising the final execution of the project seems ultimately only a minor effort compared to the preparations. Projects like this are expensive and require economic planning and budgeting quite different from traditional sculpture or painting. *Time Landscape*'s total costs (until 1978) were $74,700, which is relatively cheap compared to urban design and landscaping projects now.[247] Alan Sonfist sollicited support from the local planning board in Greenwich Village where he lived, from city government officials, the New York Horticultural

Society, and several New York banks. Now the forest's maple and oak trees have grown 25 feet tall, and shrubs and wildflowers flourish and bloom. Maintenance, often a major obstacle in such public works, especially in obtaining permission for planting at all, is now taken care of by community volunteers. *Time Landscape* has recently been acquired to become part of New York City's public art collection.

Since then, Alan Sonfist has gradually developed a number of similar time landscapes. The projects are often part of an ongoing series of activities which may stretch over years, or original ideas that are changed and re-used later, making an indication of exact dates very difficult.[248] How long and how complicated the procedures sometimes are, is shown by the example of *Monuments of Texas*, which he began at least fifteen years ago. The concept proposed a series of islands in the polluted flood plain of the Trinity River, and upon each island the artist envisioned an ecological environment, here too based upon the vegetation of the past. Part of the project is finally under way.

The issues which the artist has dealt with over the years have continuously become more complex, as have the situations he works in, including the total environment of the site, as well as the economic, social and political factors present. Among recent projects is *Sewanee Oasis* for St. Andrew's-Sewanee School (Nashville, Tennessee, 1989), resulting from a workshop/seminar. *Sewanee Oasis* is triangular in shape, within which seedlings from native species that are considered endangered were planted. The artist hopes that it will become a complex microcosm of the original forest. The concept of Sonfist's time landscapes is gradually reaching the European Continent. In Italy he obtained a large commission for a private estate in Fattoria di Celle near Florence. More important, though, is a commission by the French Ministry of Culture to create a 10 mile long series of parks in Paris, along a corridor north of La Defence, *Natural/Cultural History of Paris: A Narrative Environmental Landscape*. Inspired by some of France's most famous architecture, he decided to include references to the Chartres and Notre Dame cathedrals, the Louvre and Chaillot Palace, among others. For this he developed low platforms shaped according to the groundplan of these monuments, which will be planted with the original vegetation of the location where the buildings were constructed. This includes a pure beech forest, a mixed forest, grasslands and marshes. Also, the original topography was researched. For the terrain around the groundplans he designed floral garden patterns which refer to Napoleon's army formations during different battles. The time landscape remains a continuing thread in this project, too; new is the artist's reference to and inclusion of historical architectural landmarks. As the title suggests, natural and cultural history are connected in time. They have a story to tell, which is to be remembered.[249]

Tools for Understanding Ecology

Underlying the notion of the interdependence of mankind and nature runs the theme of survival, and the artist's deep concerns about this. As indicated above, it was one reason for him to leave the 'narrow' boundaries of the art world.[250] But, if nature is his main theme, Sonfist does not see nature's issues and technological ones as a dichotomy, as opposites. Although he aligned himself with the ecology movement, he is certainly not anti technology, recognizing the contributions that both science and technology can make to art.[251] The fellowships at the Center for Advanced Visual Studies allowed him to interact with scientists and learn about their world of ideas directly, and he has maintained this interest throughout his career. His public works to date have always involved the assistance of scientifically and technically knowledgeable professionals. "Technology can even visualize aspects of nature outside the range of the human eye, such as public outdoor projections of telescopic observations: public monuments of the sky. Many aspects of technology that now allow individuals to gain understanding of nature can be adjusted to a public scale," he wrote in his book *Natural Phenomena as Public Monuments*, a 'manifesto' of his theory of public monuments.[252]

During the seventies, he withdrew more and more from the art world to create his environmental and ecological projects. As a consequence, he received little attention from the art world for a while, with the exception of an exhibition in Florida in which Helen and Newton Harrison also participated. Meanwhile his knowledge of eco-systems, biology, and landscaping, together with his very specific approach to a site, began to draw the attention of those who wanted something different than just another piece of landscape gardening. Although he moved into the domain of the landscape architect and has become a competitor in a sense, he does not truly feel that he is a threat, but rather that his work might have a positive influence and possibly lead to a return of a concept in landscaping and gardening that is not a thin overlay to cover the natural conditions and, even worse, destroys the site's inherent environmental qualities. Meanwhile, teaching about his concerns has also become an important part of his work, and Sonfist accepts invitations to give seminars and to work with students on ecological development projects.

HELEN MAYER HARRISON - NEWTON HARRISON: The Ecological Argument

A Self-generating Eco-system

Although Newton Harrison taught painting at the University of San Diego since 1967, his interest soon switched to experiments with forms of colored light, such as glow discharge tubes, to see if light could behave like color. Thus, when he was invited to participate in the *Art and Technology Program* of the Los Angeles County Museum of Art in 1969, he submitted a proposal on "light as color in space."[253] Because the program set out to bring artists in touch with scientists, he was put in touch with Dr. Robert Meghreblian and some of the other plasma experts from the Jet Propulsion Laboratories (JPL) in Pasadena, California. After this meeting "evolved (into) a productive problem solving situation," a rapport was established between Harrison and JPL staff which existed throughout the collaboration. Harrison's research resulted in an installation of five plexiglass tubes, made by a local plastics firm, containing liquid crystals, which would change color under heat and pressure. Newton described the final effect: "...in the first tube I put an arc that was a mixture of helium and argon. The helium helped the arc path; the argon guaranteed that it would be a shocking pink-violet arc. We set it up so that the gas was injected in such a way that it started out as lightning, staying lightning for about two minutes; became an arc; stayed the arc for about three minutes; became a glow - a total glow in the tube ... the glow started to break down into platelets and then I shot more gas in so it would be an arc again. This was a ten minute cycle."[254]

It is rarely mentioned that Newton Harrison presented yet another work in the exhibition, the *Brine Shrimp Farm* which became *Survival*

Piece #2: Notations on the Eco-system of the Western Salt Works (with the inclusion of brine shrimp). Survival Piece #2 consisted of a series of 10 x 20 foot ponds, each containing brine shrimp and algae. The shrimp ate the algae, which by producing carotene altered the salinity of the water which in turn changed color, from green to olive, to brown, to brick red, as the salinity of the water grew. Designed with the assistance of Dr. Richard Eppley from the Scripps Institute of Oceanography, the piece addressed "questions of how living organisms react to specific environments."[255] In 1970 Newton Harrison also participated in *Explorations* and he was the recipient of a grant from Experiments in Art and Technology's "Projects Outside Art" program. Possibly because of this interest in technological processes, as well as the new concern for natural growth processes Harrison's career took a major turn. He decided to abandon his studio almost as a protest against what he now considered art's narrow focus on object making. One could interpret this experiment as a key work in the artist's beginning to understand something of natural ecological growth systems. Meanwhile, he had met Helen Mayer, whose interest in growing processes may have pushed him further in this direction. The first published ecological piece was exhibited at the Howard Wise Gallery, New York in 1968: *The Slow Birth and Death of a Lily Cell.*

Reflecting upon his decision to switch from experimentation with technological processes toward processes in nature, Newton Harrison said: "I had been thinking about the issue of survival as subject matter. My first response was to ponder earth - literally the ground I stood on. I asked myself what earth meant to me, and what I knew about it. And, to find out, I decided to make earth. I gathered different kinds of manure, sewage, sawdust, vegetable matter, clay and sand ... this was in 1970." And Helen Harrison commented: "And I began to utilize the piles of earth by containing them and then growing things. So, openers in earth art for us was to make earth, to plant, to grow, to harvest, to do it consciously, to change ourselves thereby. We did not consider it either interesting or valuable to use earth to make forms on unusual sites."[256] These considerations led to the development of their so called *Survival Pieces* (1971-1973). Because of their ecological concerns, these works already differed from those of other artists who worked with earth as material. Meanwhile the Harrisons had begun their collaboration as artists; a collaboration that continues until the present day.

Newton Harrison was born in New York City, 1932. He attended Antioch College, Yellow Springs, Ohio, and subsequently entered the Pennsylvania Academy of Fine Arts in 1952 where he studied sculpture. He graduated from Yale University, New Haven. Helen Mayer Harrison was born in New York City, 1929. She studied at Cornell University, New York State and Queens College, New York where she graduated in English literature. It was followed by a Master of Arts degree in education at New York University. Between 1949 and 1953 she held various teaching positions. She met Newton Harrison, and they married in 1953. In 1967, they moved to La Jolla, California, where Newton Harrison had obtained a teaching position at the University of California, San Diego. Helen Mayer Harrison obtained a doctorate degree in the philosophy of education and became Director of Educational Programs at the University Extension, from 1968-1973. Both have since then shared a professorship in the Visual Arts Department, and were Head of the Department, until their retirement. The Harrisons still live and work near San Diego. In 1993 they formed The Harrison Studio with architectural designers Gabriel Harrison and Vera Westergaard.

The consecutive *Survival Pieces*, six in total, are experiments with the self-generative aspects of different kinds of growing processes. Plants were grown in a portable pasture. Another consisted in the breeding and 'harvesting' of catfish. To create a portable orchard, eighteen trees, each in a cubic yard of earth, were placed to grow under artificial light. And as a comment on the snail infestation in California at the time, the Harrisons put white Peking Ducks in the exhibition garden, which ate the snails. Unfortunately the ducks also proceeded to eat from the real garden, which was not foreseen. The objective of these experiments was to (re)create eco-systems, in a kind of artificial setting. *Survival Piece #3: Portable Fish Farm,* exhibited at the Hayward Gallery in London (1971), probably became the most remembered of the *Survival Pieces*. Harrison's plan to electrocute the fish in public had reached the news media, causing great bewilderment and hostility toward the artist. According to the artists, this was a much more humane way of killing fish than the techniques usally applied (which was correct). But Jonathan Benthall criticized *Portable Fish Farm* for being rather contradictory in its intentions, pointing out, correctly in my view, the artist's naiveté in not having completely thought the implications of the piece through. After all, argued Benthall, "the farm requires an elaborate support system of water-heaters, agitators, syphons, etc., powered by electric current which was presumably generated at one of the big power stations in South London that belch smoke into the air. So not only did Harrison fail to recognize the literary symbolism of electrocuting fish: he also failed to recognize that its consumption of power was polluting the air of London and using up fossile fuels..."[257] Yet Benthall's conclusion was that the project was interesting and had to be taken seriously. The *Survival Pieces* presented the visitor with the process of closed (eco)-systems as art 'objects' without an aesthetic beautification. Using relatively simple technologies, the artists' farming system dealt with self-generating life-cycles. The results were obtained from detailed research, but were not meant to be used scientifically. They only served the arts. The meaning of the works consisted in visualizing ecological processes, not just growth processes. In this respect they belong to the first ecological art works.

Their experiments and problems with brine shrimp and catfish (*Survival Piece #2* and *#3*) had brought them in touch with scientists from the Scripps Institute of Oceanography. It was John Isaacs of the Institute who directed them to their research on the *Scylla errata Forskal*, a fast-growing, edible, but cannibalistic crab, which had become a threatened food source in Sri Lanka. A Sri-Lankan zoologist, Ranil Senanayake, had brought some crabs, which had been sent to him by his mother, for research on aquaculture. "We put aside all art and began investigation into the crab. We decided to let the crab be our teacher."[258] Again, it tells something about their approach: accepting that natural phenomena teach, i.e. inspire them or us, rather than them or us trying to teach, i.e. dominate nature. Their research on the crabs took such a seriously scientific turn that Isaacs suggested to them to apply for a Sea Grant from the Scripps Institute, which they received in 1974.

The research into a contained and self-generating eco-system in which *Scylla errata Forskal* would survive made them realize the complexity and fragility of eco-systems in general. It took quite some time before they discovered the conditions necessary to sustain the crab's breeding cycle. It was this section of the cycle that caused scientists headaches and the problem had not been solved yet. The Harrisons carefully patterned their working method after the functioning of eco-systems, so that their work could pass as science, but the story of their research evolved into the artists' most elaborate art project to date: the *The Lagoon Cycle* (1972-1985).[259] The portable orchards, the vegetable gardens and fish farms that were intimate survival pieces installed in galleries, had gradually expanded into planning whole eco-systems.

The Lagoon Cycle

The Lagoon Cycle consists of 60 parts, and, exhibited, functions as a photomural, 360ft. long and about 8ft. tall. All the different resources have been composed into large photographic panels, each panel containing a layered collage of photographs, maps, aerial and satellite photographs, drawings, blueprints, texts. The data are rewritten into a narrative, in which two principal actors - the Lagoon Maker and the Witness - exchange points of view on the functions of the 'lagoons.' It was first shown at the Johnson Gallery, Cornell University at Ithaca, New York, in 1985 and then at the Los Angeles County Museum of Art in 1988.[260] It is dedicated to the world's major estuarial lagoons, where ecological balance is in danger of being destroyed. *The Lagoon Cycle* is a fictional narrative about seven lagoons, each of which represents a phase in the process of thinking about the survival of the human species and the earth. The narrative takes place in the form of a dialogue between the Lagoon Maker on one side, who represents the position of the efficient organizer and developer of new technologies which promise progress and success, and the Witness on the other, who observes and comments on the possible consequences of the Lagoon Maker's proposals and plans. Among the ecological systems the lagoon is a system that, although quite fragile, reacts upon changes with the greatest resilience and flexibility in adaptation. The lagoon also serves as a metaphor for life, representing both the "fragility and endurance" of a survival system.[261]

The First Lagoon: The Lagoon at Upouveli recounts the visit to Sri Lanka and the search for the crab as a possible subject for a self-generating aquaculture system. It became a confrontation with the decline of the traditional ways of living, like fishing or the rice culture, through the introduction of new technologies like the tractor replacing the water buffalo.

The Second Lagoon: Sea Grant tells how they obtained the Sea Grant and their consistent research into living habitats and in particular into the mating of the *Scylla errata Forskal.* It narrates the idea of the tank as replacement for the lagoon, in which the lagoon becomes the metaphor for a bioregional system with an elaborate chain of interdependent life cycles. *The Second Lagoon* also critically addresses the supposed objectivity of scientific methodologies.

The Third Lagoon: The House of Crabs tells how the Harrisons are approached by "a marketing agent, businessman, scientist, journalist, and accountant," who advise them how to capitalize on the success story of the crab's breeding cycle. The story becomes an observation about the capitalist free market economy, which is one of the major causes of the discrepancies between the first and third worlds. It becomes a criticism on all existing models.

The Fourth Lagoon: On Mixing, Mapping and Territory takes the history of the Salton Sea as a point of departure for a disastrous example of destroyed ecological systems; it proposes an estuarial polycultural farming system based on a self-sustaining chain of production or a balanced life-cycle.

In *The Fifth Lagoon: From the Salton Sea to the Pacific - From the Salton Sea to the Gulf,* the Lagoon Maker proposes to cut a channel through the mountains to the Pacific Ocean or through the Colorado River delta to the Gulf of California so that the polluted Salton Sea will be fed and cleansed by the clean ocean waters and will turn into a productive estuarial lagoon. The Witness responds but "if the polluted waters of the Salton Sea are exchanged with the Pacific or the Gulf, who will flush the Ocean?"

The Sixth Lagoon: On Metaphor and Discourse takes the whole of the Colorado River basin as its subject for a discourse on how, through the canalization of the entire river, the building of power plants and irrigation systems, and our modern life style, human intervention has completely disturbed the original environmental conditions, which eventually will turn against us.

The Seventh Lagoon: The Ring of Fire proposes a ring of aquaculture systems surrounding the Pacific Ocean which would supply the world with sufficient food whithout having to resort to the industrial and technological operations that have caused the ecological imbalances we have to confront now. It considers the world's oceans as the ultimate lagoons. "Lagoon Maker and Witness begin a search for new guiding metaphors to replace those of force and fire, as they perceive the accelerating greenhouse effect as nature's response to the millennia of force and fire. Finally they muse, the oceans will rise gracefully, but will people withdraw with equal grace?"[262]

The survival of the earth is the underlying theme in *The Lagoon Cycle*, but functions here only as one layer among a number of inter-contextual levels. Newton Harrison explained: (like a collaboration of an alga and a fungus) "...which exist in symbiosis; they help each other survive. One of them takes energy from the rock, the other takes energy from sunlight. They interact and feed each other. That's nature's response to a minimum condition. At the other end, the rain forest is nature at work in a maximum condition - the conditions of maximum available energy: water, sun and heat, where plants have even evolved that grow upon plants - all surfaces support life. Now we, like nature, also cover surfaces. And we work and account for a surface, we're taking our lesson from what we're referring to."[263] The 'Lagoons' contain a body of information and data pertaining to the 'problem' to be solved for each lagoon. The information is based on scientific data encompassing geological, oceanic, atmospheric, topographical, and surface structure schemata. In addition there is information about the geographical situation, about economic, social and political issues as well as historical expositions. One may add that the whole creation of *The Lagoon Cycle* also serves as a metaphor for the way the artists perceive their function in society as artists, comparing and connecting the process of creating the work with the processes in nature.

Mapping and Metaphors

The artists' use of maps has been compared to the cartographic traditions of the past, the Renaissance and 16th or 17th centuries. To interpret *The Lagoon Cycle* in terms of a single layered map would show a complete misunderstanding of the intentions of the makers, although the Harrisons will not deny that they may have been inspired by this tradition. Yet their goals and those of the cartographer (now and in the past) are obviously wide apart. The cartographic map is basically a reduction of three-dimensional relationships in space and time to a two-dimensional scale. For the Harrisons, though, it is a means of framing their concepts; a means for them to design a conceptual model that enables them to create a 'world' that reaches out into many different 'regions' (territories, disciplines, space and time etc.) both real and imagined. It enables them to simulate a world in which all these different regions can be perceived in relationship to one another. However, these relationships are subject to processes, and to change, depending on the way we think about the urban context, the landscape, ecology, about the survival of the world, so that their mapping requires an approach which systematically layers the research facts and data. The Harrisons' intent and method would appear to have much more in common with a contemporary kind of mapping. In cybernetics, mapping is a technique of explanation, for example, whenever a 'conceptual model' is invoked. Computer simulations of complex communicational processes also use this technique of mapping. Mapping then becomes a technique to transform input into information. In a way, *The Lagoon Cycle* is art as communication, as information carrying a message. But the meaning of its message(s) can only be deduced from reading its context(s). In *The Lagoon Cycle* these messages are (re)presented in the conversations between the two protagonists in (changing) their views of structure and content on the one hand and process and context on the other.[264] Ultimately the artists' method of contextualization portrays an attempt at a new way of thinking, away from an atomistic to a relational mode of thinking, which is expressed as follows: "It creates us while we create it, since being an incomplete system, we must supply what nature does not. When this supplying is done with awareness, our behaviour alters to become a metaphor for nature. I believe this would be as true for any community."[265]

The Harrisons have been categorized as conceptual artists, story artists, ecological artists, social critics, landscape artists, and have been linked to performance art. The preparatory scientific research remained of secondary consideration. These categorizations would in a way be correct if they did not suffer from a major flaw: they only take into account one aspect, at most two at a time. However, the explorations of growth processes, for example, were from the start concerned with a complex set of ideas. Although the artists did not theorize upon these activities in terms of a 'systems approach,' like Hans Haacke, the initial approach was very similar. I refer to Haacke's research exemplified in his *Visitors' Profiles*, which installation suffered critical attacks of presenting too didactic, too literal a picture. Criticism concerning the artists' early ecological projects was related to general accusations: that they were utopian survival proclamations, showing goodwill but not well-founded upon facts, knowledge or correct analysis, and therefore appeared somewhat dishonest as far as the aims of the ecology movement were concerned. Much of the ecologically oriented art was indeed often accompanied by a rhetoric that was not sustained by the work itself, but one can hardly accuse the work of the Harrisons of such superficiality.[266] Several years later, when the Harrisons had moved part of their activities into the 'real world,' similar criticism kept being brought up, accusing the artists of having neglected the communities they were addressing.[267] The artists may indeed not have fully recognized all the political and social implications then. However, when Helen and Newton Harrison now operate outside the museum or gallery context as teachers or organizers, as facilitators offering knowledge and understanding of difficult environmental issues, their work shows the strength it implicitly holds. It is important to keep in mind that the Harrisons go about their projects as researchers set out to solve a problem, and initially they worked mainly with scientists or engineers. Yet they perceive themselves as storytellers and their work as art, whose the final form serves to transcend the literal process (in contrast to Haacke who has purposively emphasized the literalness of his work) onto a metaphorical level: the workings of the small eco-systems which were set up and in fact controlled by human intervention were in the end references to the earth's ecological systems at large. Through understanding a simple system, the Harrisons sought to understand, and create understanding about the larger, more complex systems of life. What were at first personal explorations in search of a new art form, developed into large-scale interdisciplinary research projects into global survival plans, with proposals, collaborative actions and political and social discourse.

A Different Function for Artists

The road which the Harrisons have mapped out for themselves has taken them to places few artists dare to go. Their interdisciplinary projects are now no longer only with scientists and professionals of other disciplines, but involve city planners, community groups, politicians, and so forth. It has added another dimension to their work as well as to their function as artists,.whether one calls it social work, political activism, city planning, or art consulting. *The Serpentine Lattice*, a conceptual design offering options for reforestation of the Pacific Northwest (1992-1993), for example, brought together not only biologists, ecologists, environmentalist and art historians from the university, but also local citizens and activists, former loggers and lumberjacks.[268] Their proposals now receive serious attention. *Meditations on the Condition of the Sacramento River, the Delta and the Bays at San Francisco* (1976-1977) was one of the earliest projects of such scale. Its revised version, the *Sacramento Meditations* (1980), consists of nine texts, nine mappings of California, and includes the state and federal power plants, and a scheme of irrigable and potentially irrigable land. The subject of the *Sacramento Meditations* is to make the public aware of the disastrous consequences which the enormous water consumption by the urban population and agricultural irrigation systems might have. They describe how ingenious technological constructions have reversed the flow of the San Joaquin River and the ecological consequences for the flora and fauna surrounding the original course of the river, and not only that. They narrate how a system of dams and power stations indeed regulates the the water supply of the whole central valley in California, they ask: "what if all that irrigated farming was not necessary," for instance.

The works of the Harrisons are not about presenting solutions only. More important is the call for change; to inspire a way of thinking which changes the attitude that has led to many of the problems today. Important for them is that we begin asking questions, start a discussion, a conversation. For they are painfully aware that nothing will happen if our thinking does not change. Newton Harrison: "Therefore, new paradigms will be needed which will lead to new legal and social codes that will permit land and water to be passed on to succeeding generations intact, non-renewable resources husbanded, and renewable resources not depleted. ... For if the paradigms that inform the present use and energy practices of our culture (exploit / consume / transform into goods / transform into profit) as typified by our use of the Sacramento-San Joaquin watershed do not undergo modifications slowly (through civil means) or more rapidly (through revolutionary means), then they will surely undergo modification through massive biological revolt as ecosystems simplify in response to increasing stress and become minimally productive."[269]

The following descriptions give somewhat of an idea of how the Harrisons proceed when they receive an invitation. *Guadelupe Meander: A Refugia for San Jose* (1983) drew attention to the Guadelupe River, which, wrote the artists in a letter to the Mayor and the City Council, "meanders neglected by the outskirts of your city center, almost forgotten, pursued by development on all sides, overgrown, polluted, dammed at its headwater." They suggested dredging, rebuilding of the bridges so that the river would be accessible again, the restoration of the riverbanks for native plants, shrubs, flowers to grow, and the creation of a path along the borders, so that the river would become 'the green spine' for San Jose. The art exhibition resulted in documentation of their research through photographs, maps, drawings, texts and the actual proposal was received favorably.[270]

Sometimes, unforseen events prevent their projects from happening. In the case of *Breathing Space for the Sava River / Atempause für den Save Fluß: Die Summe seiner Geschichte* (1988-1991) in former Yugoslavia it was the war. While in Berlin on a DAAD (Deutscher Akademischer Austausch Dienst) fellowship, they were invited by Hartmut Ern of Berlin's Botanical Gardens and Martin Schneider-Jacoby from the Stiftung Europäisches Naturerbe to look into the creation of a nature reserve in the northern part of Yugoslavia. The idea of creating a reserve for them still belonged to the traditional way of thinking in parts. Many of the natural reserves are the result of this kind of thinking as if it concerns islands separated from the rest of the environment. There may be small bioregional eco-systems that can exist locally, but more often the environmental conditions of the surrounding areas affect flora and fauna. In this case, the 'park' belonged to the flood plain of the Sava River, a river heavily polluted by the wastes of chemical plants, an atomic power plant, sewage systems and so forth. The artists therefore suggested a "nature corridor that would run the length of the river from its twin beginnings above Ljubljana to its ending in Beograd, where it joins the Danube River and supplies the lower Danube with one third of its water."[271] After many "conversations" back and forth the proposal was seriously considered by the Croatian Water Department and other authorities until the war, which damaged even the existing section of the nature reserve.[272]

For the Kunst- und Ausstellungshalle of the Federal Republic of Germany in Bonn, they have been working on *Future Garden: Die gefährdeten Wiesen Europas*. (1996) Five kinds of meadows which were in danger of being destroyed were moved to the roof of the building, where they will grow the next two years. After their work on the threatening deforestation in North America the Harrisons became interested in what happened in Europe. They were told that the problem here was of a completely different nature: the disappearance of the meadows in Europe; some of them centuries old and belonging to the richest existing biodiverse eco-systems. And they arrived at the conclusion that the European meadow developed from an unconscious collaboration between culture and nature. And that it appeared one of the last examples of biodiversity on the whole continent. They visited numerous meadows and found a 400 hundred year old one in the Eifel which existence was threatened by a housing project. It was decided to try and save it and bring it to the roof of the Kunst- und Ausstellungshalle. This was the beginning of a project which led to the creation of the "Meadow Stories." Herein the Harrisons proceed their discourse on biodiversity with warnings, possibilities and choices for future generations to preserve and develop meadows and gardens to show respect for the original local vegetation.[273] For not only the disappearance of the meadows is at stake; their destruction will effect complete water retention systems and ultimately endanger our drinking water.

Another large project still in progress is *Tibet is the High Ground,* in which the Harrisons propose reforestation of the Tibetan Plateau. Using photograhs, map drawings, poems and performances, they present the idea of an "analog" forest, or "a simplified woodland-rainforest eco-system," which will be less complex than the original, but still bring yields to its inhabitants, by protecting the existing topsoils and the riverine and wetlands ecologies along the rivers that originate here. And these are the seven largest rivers of the South-Asian continent: Indus, Yellow River, Yangtzekiang, Salween, Mekong, Brahmaputra and Ganges. The woods of the Tibetan Plateau have been systematically cut and carried off, in particular since the Chinese invasion after the Second World War. How grave the consequences of this destruction will be has not fully been comprehended. Erosion and floods are common features now. But, ask the Harrisons, "Can it be true that (even) the drought in California and inland Oregon is a result of the northward shift of the jet stream across the Pacific so that the moisture laden air instead gifts its waters to the Pacific northwest? Can it be true that the jet stream is veering south to avoid the chimney effect over Tibet? Can it be true that the chimney effect over Tibet and the emerging desert is a response to the loss of forest and other vegetational cover?"[274]

Formally the work of the Harrisons has not changed much. Mapping and storytelling remain their footholds, both being central elements in the long human tradition of knowledge and information transference. Newton Harrison has also described their method as "conversational drift." "When we get up to tell stories (to the Croatian water Department), the storytelling is what causes the conversation to drift. After we've told our stories those government officials see our images differently. The conversation drifts and in this context the idea of purifying the whole (Sava) river seems more real and less difficult." Graig Adcock interprets the work of the Harrisons: "By doing art with ecological content, the Harrisons imply that the human species should treat the planet as sculpture. Such an idea strikes many people as absurd, but humans are clearly modifying the ecosystem and changing the fragile biosphere of the planet. The Earth will become and already is very largely, an artificial construct."[275]

However, rather than submitting to a doomsday scenario they maintain an optimistic attitude. The principal argument in their strategy against entropy is the ecological argument, and their scientific research withstands comparison with environmental impact studies. Questioning certain scientific and technological methodologies, they suggest that nature, science and technology are not incompatible. Probing the boundaries of art with non-artistic 'scientific' methods, they bring these disciplines into the domain of art. Although they question the modernist concept of art, they perceive themselves as artists in search of a new territory for the visual arts. Although their work has been accused of using traditional means, only a superficial interpretation can lead to this conclusion. The layering of the work comprises information of scientific data, historical schemata, geological processes of the past and present, information about natural landscape conditions and human involvement, presented as interrelated events that have evolved over a period of time and space. There is an unmistakable concern about current and future ecological disasters, using metaphors as a strategy to voice their warnings and suggestions for change. Their message: the need for everyone to take responsiblity for their environment.[276]

Chapter 9

Toward an Art in Public Places

Although one cannot say that the current explosion of art in public places would not have taken place without the instigation of artists like Turrell, Irwin, Sonfist or the Harrisons, their exploration of a new function for sculpture stands at the beginning of this growth, and has certainly inspired a 'second generation' of artists now working in the field. At the same time we witness the development of an infrastructure of arts commissions, institutions and organizers specializing in this field. During the eighties more and more innovative arts agencies and city administrations began to link municipal development and arts politics. Not only more and more states pass legislations requiring a mandatory percentage of public building to be spent on an art project, but in addition communities are developing policies and procedures which exactly prescribe how the allocated funds are to be spent. New funding opportunities, federal, state, local and private or corporate, have established themselves. One more aspect needs to be mentioned. By now, the term Art in Public Places has become a familiar word with a large group of artists.[277] The development of a whole network, not only involving arts organizers and administrators, but many different professionals outside the visual arts, has also led to a whole new realm of structures and issues for the artists to familiarize themselves with if they seek to work within this context.

The kind of art that results from it has to answer to a new set of criteria and qualities. The starting point is that the work has a relationship with its environment. Its purpose is to make a contribution in shaping the public space, whether urban or 'natural.' In order to get away from the sculptural object that is just plunked down, cities have set up arts commissions and introduced procedural policies which require that artists, architects, landscape architects and designers collaborate from the start. The 'old' artist was commissioned to create an object. Robert Irwin said: "...only marginally public generally (it is) consigned to an out-of-the-way place, a leftover cul-de-sac or a ceremonial spot of the architect's choosing.[278] The 'new' artist is involved in the design of a whole park or plaza. The idea of sculpture as object is changed in favor of a process of collaboration with architects, designers, engineers, environmentalists and local community groups, with a view to making the public space itself a work of art. The old public art object is transformed into an art in public places. The history of many of Irwin's projects gave an indication of the tasks that lay ahead of an artist who decided to move into this field, whether in the form of an environmentally oriented work of art, or landscaping and land reclamation projects. The projects of. Bulgarian-born Christo are another good example. The proposals for his temporary wrapping projects like *Valley Curtain* (Grand Hogback, Rifle, Colorado, 1971-72) or *Running Fence* (Sonoma County, California, 1976) were generally not uncontested. Christo had to negotiate with politicians, discuss with community groups and environmentalists, or obtain permits from authorities. For *Running Fence* he was even asked to file an Environmental Impact Report, guaranteeing that the site would be left undisturbed and returned to its original state. For the actual realization, he has to hire specialists and a large crew of workmen. Contrary to Robert Irwin, who perceives these aspects as necessary evils, as a part of the job like preparing a canvas, Christo considers all these things as integral part of his work. In this respect, Christo is still connected to the 'art is life, or life is art idea' that Joseph Beuys represented.

The discussion of the development and construction procedures of such an 'earthwork' shows how different the function of the artist has become from the studio sculptor or painter. Not only has the artist an obligation to respond to the context and acquire the necessary knowledge, but he or she also has to understand the shift in working procedures involved in this type of work. In general, the following procedures are followed. First the artist is invited and/or submits a proposal, including detailed descriptions, plans and budget information, which is then juried by 'specialists in the field.' If the artist is selected he or she will generally have to sign contracts and sometimes be prepared to account for the implementation of the project with detailed reports. The artist has to be prepared to deal with government agencies, environmentalists, and the public. He or she has be prepared to have the proposal subjected to feasibility and environmental impact surveys, and to work in a team of professionals, as a project developer or supervisor. Obviously not every artist feels equipped for this task. It takes another kind of personality than that of the studio artist. Los Angeles-based sculptor Lloyd Hamrol described this position in an interview: "Yes. Dealing with public projects is complex, and sometimes very frustrating. The second part is that it often takes years to complete a large project; there are delays, interruptions, and money complications. ... It's a world of practicalities, of economics, and doesn't have anything to do with the language of forms and materials that I use. ... The artist encounters people on technical and administrative levels, and not just in their professional roles, but also who they are as individuals. I see my role as planner and orchestrator of my projects as an outgrowth of my teaching experience."[279]

Since the beginning of outdoor environmental works in the mid-seventies, directions in public art have become quite diverse. Artists have become involved in recultivation of landfills and wetlands, and land reclamation projects, in the design of urban landscapes and parks. I have distinguished three directions based not so much on formal aspects or similarities as on intent and purpose: land reclamation, ecological art, and urban design and landscaping.

Land Reclamation

If Robert Smithson remained unsuccessful with the strip-mining companies in claiming support for his land reclamation proposals in 1972, half a decade later the idea of land recamation as public sculpture was receiving growing attention. Interest came from different sources. The interest of redevelopment agencies began in the late seventies. By now most major U.S. cities have established an agency dedicated to the redevelopment of designated sites. (The word redevelopment may or may not be part of the agency's official title.) The restoration of land for public use is only a small part of the flurry of activities involving a recuperation of nature. Artistic involvement in projects of mining companies, which are in fact subject to strict regulations to restore the land after (ab)use, has remained small but is growing. The majority of projects in this area are still initiated by art institutions, or programs already involved in the arts, rather than emanating from the companies themselves.[280]

The initiative of King County Arts Commission in Seattle, Washington, was among the first land reclamation programs. Visual arts coordinator Jerry Allen wrote that the intention of the *Earthworks:*

Land Reclamation as Sculpture project was "to create a new tool in the rehabilitation of technically abused land, by giving artists the opportunity to design for surplus gravel pits, surface mines and landfill sites. In King County, like most jurisdictions, the problem is acute, with 104 publicly owned and many more privately-owned surface mine sites. Few of these sites have meaningful reclamation plans. Thus they are scars that dot the landscape."[281] The development of the program took two years (1977-1979), before one could even think of commencing the execution of one of the prososals, because the most important task for the Arts Commission was the building of a support system, and the allocation of sufficient funds from various non-art organizations.

From the artists who had been invited to send in a proposal, Robert Morris was chosen by a selected jury to create an artwork at Johnson Pit #30 of the same title. Robert Morris' construction notes in the technical report give week-to-week information about the process and procedures of the undertaking. He describes the preparatory, executive, and management procedures. The preparatory procedures, for example, included hydraulics, soil investigation, an environmental impact assessment, and a bidding of the job to find a suitable contractor, as well as establishing procedures for the construction phase, such as soil testing, site drainage, hydroseeding, excavation, etc. The King County Arts Commission had even appointed a project assistant especially to supervise the procedures.[282]

Of the other proposals, only Herbert Bayer's *Mill Creek Canyon Earthworks* (1982) was built. Bayer created a park, covering 2.5 square acres, composed of geometric elements which follow the contours of the land. He designed two mounds, a round one and an oblong one, a mound with a ring and a pond, and a cone with a bridge over the creek which runs through the park. The area being a retention basin, the mounds function as drainage as well as seating for visitors. The design and the construction of the 'earthwork' was in fact only a part of the whole reclamation project, which also included the construction and maintenance of the retention basin. This part, however, fell under the responsibility of the City Engineering Department, costing $450,000.[283]

Earthworks: Land Reclamation As Sculpture shows the new dimensions involved in the development and implementation of such a program, even if it seems relatively simple, since the projects chosen mainly involved reshaping the dirt of the site. Robert Morris accentuated the contours of the pit into a kind of inverted terrace. For Herbert Bayer the functional aspect was a major consideration in the final design. In addition, he used the circular shape of the earth mound for cone and ring mounds situated in the canyon's landscape to integrate the existing and newly built parts of the landscape with as little interference as possible. Although both artists used formal references to historical and architectural concepts, those being integrated in the existing elements of the sites, the works became an integral aspect of the landscape, and subject to its continuous processes of change, natural or man-made.

Although Robert Morris was concerned that artists might be easily exploited because they tend to work more cheaply than commercial reclamation companies, more artists are invited to develop concepts for landfills and industrially damaged sites or quarries. Even though artists may not be accepted as equal partners yet, the successful examples of finished projects like Nancy Holt's *Sky Mound* or Harriet Feigenbaum's *Erosion and Sedimentation Plan for Red Ash and Coal Silt Area - Willow Rings* (near Scranton, Pennsylvania) should make us aware of the yet untapped potential of artists' creativity. Planted in 1985, the three circles of sixty willow trees surrounding a pond have continued to grow and cover the once wasted 15 acre site of a coal strip mine. It is now part of a wetlands wildlife preserve.

Ecological Art

Land reclamation projects are often combined with ecological concerns and the two often overlap. Projects of the late eighties and nineties show how ecological awareness has become a dominant theme. Since Alan Sonfist and the Harrisons have started their ecological art works, another generation of artists feels strongly obliged to make a contribution, even if it seems like carrying coals to Newcastle. Everybody knows that we need to reduce our energy consumption drastically, whether it is oil, water or wood. The oil reserves are predicted only to last another forty years. Water will become so precious, that we will hardly be able to pay for it. And what will happen to the atmosphere if we continue to log the rain forests?

Patricia Johanson's work actively intervenes and intentionally transforms sites into biodiverse systems in respect of the geographical and climatic conditions. She perceives her sculptures as "habitats" where flora and fauna can thrive as an organic interdependent system to revitalize and purify degraded environments. About her *Leonhardt Lagoon* (Dallas, Texas, 1981-1986) she writes: "The lagoon was in the middle of Dallas' largest park with four major museums along the shore, and it seemed a wonderful opportunity to convert it into a home for native wildlife - ducks, turtles, fish, shrimp, insects - by cleaning up the water and conceiving of landscaping as food. The 'sculpture' was thought of as not just aesthetic, but rather a means of bringing people into contact with the plants and animals and the water."[284] She also reflects on previous utilizations. In the case of *Endangered Garden*, conceived for Candlestick Cove on the San Francisco Bay between the ocean and highway 101 (1985 - present), it was the ancient native American culture. The goal of this project is to revitalize the original biodiversity. Johanson's design encompasses a sculptural form in the shape of a ribbon worm, called *Ribbon Worm Tidal Steps,* the planting of native vines and shrubs, and the (re-)creation of birdpaths, and shellfish habitats, referring to the original shellfish mounds here, which were not constructed, though. Pathways allow the public to walk down from the baywalk above and enter the wetlands, depending on the tidal waves. As a delicate wetland, it is subject to the regular rise and fall of the tides and has a complex eco-system. The artist wants the visitor to explore this superficially uneventful situation, which on close inspection reveals an intense and lively animal population, in order to make the public aware of the extraordinary natural phenomena right in front of their eyes. This educational aspect contributes significantly to the meaning of her work.

Another ecological landfill project in the same region is Agnes Denes's *North Waterfront Park.* Together with landscape architects R. Haag and J.N. Roberts she developed a large land reclamation project which is being realized for the city of Berkeley, California. The *North Waterfront Park Master Plan* (1989-1991, date of approval), a landfill situated on the eastern shore of the San Francisco Bay, is designed to transform the 97 acre site into a park with a self-regulating bioregional eco-system. The designers did not want to create an environmentally sound but isolated eco-island, but started from a concept which allows room for growth and change, even normal erosion and decay, thus functioning as an open-ended system. Denes describes the proposal as follows: "The overall organization of the park creates a time warp, going from civilization (entrance) to wilderness (end). Walkways and paths crisscross the park, offering changing views as they wind around mounds and trees. A perimeter path circles the park offering views and access to water. ... A 12 acre wildlife sanctuary establishes the whole park as a life-harboring environment for an increased number of species, and bioremediation programs offer techniques for cleansing landfill pollutants."[285] Agnes Denes's work has been scientifically and interdisciplinarily oriented since the beginning of her artistic career in the sixties. Especially mathematics has been of interest to her, and is a recurring element. Since 1992 she has been working on a commission with other artists to develop art concepts for the Wildlife Preserve and Research Center in Columbus, Ohio, called *The Wilds*. Its purpose is again to reverse processes of erosion, and the extinction of regional flora and fauna. "My concern is with the creation of a language of perception that allows the flow of information among alien systems and disciplines, eliminating the boundaries of art in order to make new associations and valid analogies possible," writes Denes.[286]

Urban Design and Landscaping

The rise of environmental and ecological concerns coincided with the rapid deterioration of the inner cities, growing crime rates, traffic congestion and air pollution. This recognition had at least one major consequence: it has led to concepts in urban development and design which take into account the community and even the whole city. The loosely thrown together megastructures and glass towers of the sixties had no connection to the outdoors. One entered via the parking structure. To counter this trend, another 'movement' began with rehabilitation programs, which looked into relationships with the community and the surrounding architecture, for example. In the mean time it became accepted that art can and does contribute in the widest sense to enhancing the quality of life through beautification or may contribute to an historical consciousness in the commmunity. Politicians came to the conclusion that cultural facilities are economically viable. With this in mind, the introduction of percent-for-art and other programs in which artists are invited to work with architects and landscape designers can be explained more easily. In this view, it no longer suffices to design a building as a single monument that carries only the stamp of the architect.

Since the eighties, a number of cities have introduced outstanding progressive art programming policies that consider the visual arts an integral part of the network of social and political activities. Within these programs, artists and architects now collaborate on buildings, landscaping, and even public transportation systems. The City of Seattle led the way in the practical integration of art within communal planning. Its Arts Commission is generally credited with being the first to explore the use of interdisciplinary design teams on contemporary art projects since 1983. Involving artists from the beginning of the development of a public building has become accepted. Los Angeles followed with a similar concept, developed by the Los Angeles Community Redevelopment Agency (CRA) in 1985, when the first *Public Art Policy* was introduced, setting official guidelines for the implementation of the percent-for-art legislation. The focus here was on social context and community participation in respect of the city's multi-cultural heritage, seeking civic responsibility and spirit. Local and regional artists were to play a central role. This policy intended to answer a major issue of today: to try and repair some of the damage done to the daily environment, "to eliminate blight and revitalize the citiy through redevelopment in designated residential, industrial and commercial areas" with the sole purpose "that these activities have a major impact on the citiy's design and aesthetics, its livability and workability" by bringing together the revised ideas on "urban landscaping" and creating a function for the visual artists to participate in this process.[287] Since its inception the Community Redevelopment Agency has fostered a series of public art projects for redevelopments in downtown Los Angeles, such as South Hope Park, a 2.5 acre landscaped park, and the redesign of Pershing Square. All these projects aim at reshaping the downtown area, so that it may regain (some of) its old lustre. Whether this is at all possible, considering such insurmountable problems, as homelessness and crime, remains to be seen; the attempt is admirable. The Los Angeles County Metropolitan Transportation Commission has developed a similar program for its new metro rail system, called Art for Rail Transit, in which artists work with architects on the design of the subway stations.[288]

An outstanding example is *Battery Park City Plaza* in New York, a collaboration between artists Scott Burton and Siah Armajani and architect Cesar Pelli. Burton and Armajani were selected from a competition to collaborate with Cesar Pelli and landscape architect M. Paul Friedberg to shape the plaza surrounding the World Financial Center, designed by Pelli on the tip of Lower Manhattan, in Battery Park City (1982-1989). Initially, Pelli was not enthusiastic at all, and the Arts Committee seems to have had a difficult task convincing him. "I thought it was a terrible idea," he said in an interview. "My worry was that I'd be saddled with an artist who would be primarily interested in expressing personal and esoteric pursuits. There are many examples of anti-civic art, where the building is seen as background for the artist who feels he must make a confrontational point."[289] Pelli's reply shows that there exist quite a few misconceptions between architects and artists regarding each other's professions. However, in the end he was convinced, and the result was a true collaboration, where the individual ego was subordinated to the larger philosophy of the future function of the site. For Siah Armajani, who had worked on other collaborations, the task was not new. He had already stated his point of view about art in public places: "Public art's immediate concern is not with the artistic but with the work it is meant to perform: it should be open, available, useful, common and near public spaces."[290] The design team divided Battery Park City Plaza into four principal design areas, connected by walkways and stairs: a broad paved plaza, waterside seating, a small green park and a garden with labyrinth and fountains. There are enclosed, private areas, where one can sit quietly, and spaces that resemble a marketplace, where people meet for lunch, do business, etc. The design combines structurally laid out functional-sculptural elements with artificial-natural grounds that may not offer an encounter with an impressive work of art, but try to answer to personal needs of a particular moment. Although the project is already a few years old, it is considered one of the most successful in its kind. As indicated before, this approach not only poses a different conception of art, its also puts forward an artistic role which no longer emphasizes individual achievement. It is not the autonomous art work any more that counts. The art work becomes subordinated to the requirements of the larger 'whole,' and the final result may not even be recognizable as art. It is obvious that the interdisciplinary collaboration between different disciplines implies a change in production and working methods, in which, ideally, there is a professional exchange of ideas on an equal footing. And ideally, the result will be a work whose existence is justified by its relationship to the environment as a whole, its multiple social and cultural layers. The final purpose of such a project, its service to the public, comes before the artist's personal aesthetic, even. The possibilities for artists to work on urban design and landscaping projects have continued to grow in the nineties, offering opportunities for a new kind of art in public places.

PART III

Chapter 10

Nam June Paik: Media Visions

Early Electronic Media

After 1965, the year in which Sony Corporation introduced the first portable camera/recorder, making it economically feasible for artists to use video in their work, artists from many different fields - sculpture, performance, photography, etc. - began making videotapes. These early works were primarily restricted to experiments with the technical features of the medium, such as video feedback and closed-circuit. The possibility to view the images filmed just a few seconds later allowed for an immediate reflection. This made it attractive for artists involved with dance or music, giving them instantaneous real time reference to their movements, The results were directly visible on the monitor, so that video became just the right medium to complement the concerns with the self and the body prevalent among performance and body artists. Bruce Nauman's, Terry Fox's or Dennis Oppenheim's early videotapes fall into this category. The literature about these technical aspects in relationship to its first explorations by performance artists abounds.

The advent of the portable video recorder was also greeted as a new tool that might change the structure of television. It could become another means to distribute information and a new educational tool. Apart from discussions comparing the new medium to film, comparisons were made between broadcast television as a one-way communication system and the videotape recorder which had input (camera and microphone), storage and processing (the record/playback deck), and output (monitor), and thus functioned as an entire information system. It allowed personal feedback, which in turn extended the mode of communication and control. In video 'feedback' is used to describe the process of returning a signal to its source, making video, as it were, instantaneous. As said, the most important technical features of video equipment are described as feedback, and real time processing, language with which we have become familiar in cybernetics and systems theory. The idea of video as a two-way communication medium principally attracted those artists who saw it as a means to influence the existing television structure. The television structure, from production to broadcast, was totally consolidated by the mid-sixties and in the hands of major networks (ABC, CBS, NBC) and a number of public stations (WGBH, KCET, WNET, KQED). The home television receiver had been introduced in the United States in 1946, and by the end of the sixties almost every second American household had a television set. Television was seen as the technology with the greatest influence on western society since World War II. It was often looked at suspiciously by the educated as a medium, an information system, which by nature or by way of its organizational structure and popular programming, was bound to brainwash the viewer. Alternative and guerrilla TV groups sprang up like mushrooms. Many artists perceived the networks as a bulwark of control and power, to be attacked in much the same way as the existing art world. Nam June Paik and Wolf Vostell demolished television sets as early as 1959, as a part of their environmental media collage pieces and Fluxus activities. The early use of video by artists who were looking for new material turned it into an anti-art statement, similar to the earth art movement. However, it was, relatively quickly recognized as means a of expression in art, and the history of the co-optation of video art is not so different from land art. Soon galleries and museums showed interest in this 'non-commercial' medium.[291] Few developments were followed as closely as those of the use of video in art, and by 1973 the 'history of video art' was written as one of painting and sculpture, as a medium to be compared with other technology-based media, that had infiltrated the visual arts since the nineteenth century, like photography and film.

In the beginning, writing on video was often restricted to an explanation of the technical features that qualified the videowork. For example, the late film curator David Bienstock wrote in the program notes accompanying the first videotape exhibition at the Whitney Museum of American Art (1971), which consisted almost entirely of image-processed works: "It was decided instead to limit the program to tapes which focus on the ability to create and generate 'its own intrinsic imagery,' rather than its ability to record reality. This is done with special video synthesizers, colorizers, and by utilizing many of the unique electronic properties of the medium."[292] Having no antecedents, the development of a comprehensive theory of its aesthetics caused problems, because few scholars or art writers at the time were able to deal with the medium as part of the larger social context in which television was already playing a dominant role. An exception was René Berger's distinction between a macro-, meso-, and micro-level of television. Macro-television were the networks, meso-television consisted of cable stations and the like, micro-television was the work of independent producers, including artists, and groups whose productions were created with portable equipment.[293]

A few curators like David Ross and James Harithas, who were involved with the 'video as art' phenomenon early on, were able to see the new medium as an instrument of complex layers in which the technological, the cultural / art components, the political, the institutional / television, and the economic / production all came into play. They perceived video as a system of production, distribution and reception, with none of the qualities of a precious art object. For the videotape is just a tape when not played. Secondly, the possibility of replication as an inherent quality of videotape distinguishes it from being 'unique.' In principle, it is exactly this technical feature that would have enabled the 'art tape' to partake in another existing marketing system, the audio and video cassette industry. However, most video artists have dealt with this aspect rather ambiguously. Initially intending to derange the art system with non-precious and replicable videotape, many artists still wanted their work to be treated as an art object. Copies were and are rented out for a hundred dollars or more per screening by special video art distribution centers. Occasionally, the artist will sell a master copy to a museum. Only since the late eighties the video art distribution centers have been offering cheap 1/2 inch copies for a reasonable price (ca. $65), just a little more than your popular video film at the videostore.

As technical developments proceeded, many of the 'video artists' of the first hour returned to the confines of the art world and to the medium they originally came from. Many made only a few tapes, like Richard Serra, Nancy Holt, Beryl Korot, Peter Campus, Robert Morris, or Terry Fox. Others kept returning to video occasionally, like Vito Acconci, or Bruce Nauman. There were several reasons for that. Art itself had changed directions again. Performance and body artists who experimented with video feedback and the closed-circuit format, had moved toward theatrical and professional, multi- or intermedia productions. The medium video was rarely used to document processes now, and nobody was interested any more in long hours of unedited minimalistic actions, taped without interruption. Other artists switched to the sculptural video installation, which fits the traditional concept of sculpture as object and is thus better suited for museum and gallery presentation than the videotape. The video sculpture allows the spectator to pass by or walk through the exhibition space as usual. The expectations with which the visitor looks at the work are those of an art object.[294] This development actually parallels a general pattern of the art world in which attention was paid again to the autonomous work of art. A video installation is a composition of television monitors to which an ambient frame may be added. A video installation, or video sculpture, as Wulf Herzogenrath preferred to call these works, is defined by its spatial setting.

But there was yet another development. In the seventies, a generation of video artists familiarized themselves immediately with the digital and electronic technologies. There was also a tendency to specialize again; a direction that was partially determined by the growing technological sophistication. This trend has continued until today through the introduction of better cameras, the digital computer, interactive laser discs, electronic networks, virtual reality technology, etc. The videotape still exists, but for how long? Now that video and computer works have all become digitized, they have become part of the same family of electronic digital media. And it is in these areas that the most far-reaching concepts for new art forms are have developed, for the rapid developments in information technologies have totally changed the context in which video/television and the computer operate. The complexity of the technology, even if it becomes ever more 'user friendly,' requires a constant up-date of knowledge. This means that the artist has to learn how to use 'Harry,' CAD systems, or has to hire an editor; that he has to become a hard- or software developer himself or to collaborate with a computer programmer/scientist.

Interestingly, though, a number of issues heavily debated in the sixties have come to the fore again in the art and technology discussion. One of them is related to the non-objectness of digital technology, as a consequence of which the discussion about art as a system has opened up again. For video and the computer are technologies that operate in a larger structural frame or network of communication systems.

Secondly, in dataspace there is no definite version of a work. The status of digital work depends on the history of a specific set of interactions; each interruption in the information stored will create another work in progress. In an interactive computer - video or telecommunication or satellite work of art - the final result depends on the interactions of the participants, as in an open system. This aspect will affect the traditional concepts of authorship and authenticity, something which Nam June Paik already dealt with in his earlier work, but which will become important as the digital and electronic technologies become more and more networked. It goes without saying that if an artist wants to be involved in these developments, he or she will have to move out of the art system as it functions now, far beyond the frame of the television set and the computer screen, into the real world.

The following chapters discuss the work of artists who have played a major role in the development of media art forms from the start, and still do, and who have created a new role for themselves as artists: Nam June Paik, Bill Viola, Paul Ryan, and the Vasulkas.

Nam June Paik: Crossing Boundaries

Nam June Paik started his artistic career within the Fluxus movement, and for a long time remained associated with them. Fluxus, whose roots can be traced back to Dada, wanted to reconnect art with life, and rid itself of the hierarchy that governed the art system. However, Paik's concerns went beyond the "art is life" ideals. Said Paik in retrospect that "breaking the boundaries of the arts" meant for him reaching out into "the boundary regions between various fields, and complex problems of interfacing these different media and elements, such as music and visual art, hardware and software, electronics and humanities in the classical sense..."[295]

History decided that he was to become the 'father of video art,' though, and his main body of work consists of the medium video. However, he has never allowed his work to be pinned down to strict categories and his interdisciplinary activities exemplify this. Nam June Paik was born in 1932 in Seoul, Korea. His family had to flee to Japan in 1950, because of the Korean War. Paik studied music, art-history and philosophy at the University of Tokyo, where he graduated in 1956 with a paper on Arnold Schönberg. He continued his music studies in

Munich and Freiburg, Germany. Nam June Paik's interests soon shifted to electronic music and 'action-music,' and he became involved with Fluxus. He participated in the *Fluxus - Internationale Festspiele neuester Musik*, at Wiesbaden, in 1962. It was followed by his first, much remembered exhibition at Galerie Parnass, Wuppertal, in 1963, entitled *Exposition of Music/Electronic Television.* Paik returned to Japan the following year and began his first experiments with television electronics and electromagnets together with an engineer, Shuya Abe, to develop a concept for participatory television - *Participation TV* - in which the viewer was requested to actively participate in the creation and re-creation of the images. Curator James Harithas concisely phrased Nam June Paik's position in the introduction to the first major retrospective of the artist, staged at the Everson Museum of Art, Syracuse, NY, in 1973, *Nam June Paik: Videa 'n' Videology (1959-1973)*: "Paik's esthetic position is based on broad cultural experiences. In part, his thought and work show the effect of Buddhism, his training in classical and electronic music, his involvement, his penetrating interest in the work of John Cage and Norbert Wiener, among others, and his incessant experimentation with cybernated systems and television."[296]

By 1973, Nam June Paik had already experimented with various media, such as music composition, performance, 'sculpture' and video. He had even gained a certain notoriety as a Fluxus performer (in collaboration with Charlotte Moorman). From then on, he decided to express his ideas about art and television, and about the new communications media via multi-monitor installations, videotapes, television and satellite programs. However, Nam June Paik did not stop at the crossing of the boundaries, but acquired knowledge and expertise from a number of disciplines. He experimented with technology not only out of curiosity, to make a technological art, but was just as interested in the philosophical, social and political implications of the media he explored, viz. television, video, computer applications, satellites. For Paik, television was not merely a box in our living-room or bathroom, not merely a screen presenting images. Television represented a whole new set of social and political ramifications, including production and broadcast control by the television networks. It involved questions as to who controls what is to be aired, in what format and when, as to the behavioral aspects related to viewing television and whether and how it might affect other types of social conduct. Paik informed himself about the technological developments as well as the related scientific theories. They influenced his work as much as some of the earlier art movements, like Dada. In the early sixties he met John Cage who was to influence him deeply. They became life-long friends. Cage had already introduced the notions of chance and indeterminacy into his compositions and performances, and indicated the obsolescence of the theoretical ideas underlying classical and modern music in 1961, referring to the new electronic media in a collection of essays published under the title *Silence*. Cage: "It is high time to let sounds issue in time independent of a beat in order to show a musical recognition of the necessity of time which has already been recognized on the part of broadcast communications, radio, television, not to mention the magnetic tape, not to mention travel by air, departures and arrivals from no matter what point at no matter what time, not to mention telephony."[297]

Considering Paik's research into the new electronic technologies and computer sciences, we may assume that these concepts were just as important to him as the ones in the arts. For example, in his New York archive, in a chest of drawers with 'personal junk' from 1961, Paik collected an article written on the occasion of physicist Werner Heisenberg's birthday, in which the scientist's research into the essence of matter is described.[298] As said before, in Heisenberg's model, the unobserved atom, or electronic particle, for instance, does not have a definite position, but exists only as a tendency and has the inclination to be in several possible positions at the same time, so it is not actually anywhere, but potentially everywhere; attributes that may be ascribed to it are possibilities, not actualities. Nam June Paik connected the theory of cybernetics, in which the interdisciplinary approach was a major principle, with his own experiments that crossed over to different disciplines. In his writings he regularly refers to Norbert Wiener's and Marshall McLuhan's theories in relation to his own ideas of the communication aspects of art. The article "Norbert Wiener and Marshall McLuhan" (1967) sets out to compare Norbert Wiener's theory of cybernetics with Marshall McLuhan's media theory, and quotes both at length. Paik wrote: "Twenty years ago Norbert Wiener, whose hobby was to read the Encyclopedia Britannica from A to Z, anticipated the Intermedia. ... The above conception of intermedia brought forth an interscience called cybernetics, and the latter pushed the electric age (engineering with the technique of strong current) into the electronic age (control and communication using the technique of weak current), which was explored as the escalated 'Mix Media' in Marshall McLuhan's global village. ... Norbert Wiener wrote that the information, in which a message was sent, plays the same role as the information, in which a message is not sent. It sounds almost Cagean. ... Indeterminism, a core in the thought of the twentieth century from Heisenberg via Sartre to Cage, reflected also in Wiener and McLuhan. For Wiener indeterminism was entropy, a classical terminology of statistics, and for McLuhan indeterminism was 'the cool media with low definition'. ... It is illuminating to seek the common denominator running through these parallels. ... Wiener used these characteristics as the micro-form to construct the technical interior of the electronic age, whereas McLuhan used them as the macro-form to interpret the psychological and sociological exterior of the electronic age."[299] It goes without saying that Paik's understanding of Norbert Wiener's concept of a message in terms of communications theory and Marshall McLuhan's concept of a message as information received via the media - "extensions of man," as McLuhan called them - was related to the communications aspects of his own work in electronic television, and that he understood the implications well. For it is illuminating that Paik quoted exactly those passages in which both authors discuss the essence of (a) message, when communication becomes information in the 'electronic age.' As far as references to Wiener go, Nam June Paik chose the interdisciplinary approach of cybernetics, as related to the concepts of 'intermedia' and 'crossing boundaries.' Fluxus may have provided the young artist with a taste for chance and accident, but throughout his career, the artist kept returning to the scientific definitions of indeterminacy and randomness. This is apparent from the terminology he used when answering, for example, Jud Yalkut's question regarding a creative use of the computer in art with: "The more it deals with the character of randomness and repetition, the more efficient is the computer. These are the two poles of human artistic materials. Total repetition means total determinism. Total randomness means total indeterminism. Both are mathematically simply explicable. The problem is how to use these two characters effectively. Therein lies the secret for the successful usage of the computer in the creative arts."[300] In terms of video, the notions of chance and randomness were related to qualities like low resolution or low definition, and electromagnetic disturbances, but also to the possible random effects which one could create with its feedback options and build into the digital computer when it was hooked up to the video synthesizer. It has always fascinated Paik to create seemingly (or really?) random effects with a technology apparently completely controlled.[301]

For example, a result of his early collaboration with the Japanese engineer Shuya Abe involved the construction of a robotic sculpture, *Robot K-456* (1964/5). The life-size, fragile-looking robot with its body of metal parts and wires and its motorized feet had a remote-control, a built-in audiotape with John F. Kennedy's inaugural speech, and threw white beans on command. The sculpture was generally interpreted as an extension of the kinetic tradition. But Paik's attitude toward the work differed essentially from that of the kinetic artists, who were mainly interested in formal and optical aspects of movement. Instead, Nam June Paik's concerns should be explained in terms of cybernetic notions of control and communication. It was part of his research into indeterminacy, that is into controlling the unpredictability and randomness of the robot's movements. The artist presented his robot regularly in exhibitions and performances. For the videotape *A Tribute to John Cage* (1973) Paik and Cage had their robot, guided by remote-control, wander through the streets of New York, and staged an accident with it. Therefore, one might prefer to see the semi-operative automaton as an ironic gesture toward the popularized notions of cybernetic control, Cyborg Art, as it was called by Jack Burnham, who described *Robot K-456* with its 20 Channel Radio Control and 10 Channel Data Recorder as a robot "stripped bare of everything but her skeletal aluminum components."[302]

Nam June Paik has never clearly expressed himself as one of Marshall McLuhan's adversaries nor admirers, but in effect, one of his early videotapes was a pun on the ideas of Marshall McLuhan. In *McLuhan Caged* (1967), he portrayed the media guru in the form of a caricature. In 1967, NBC and CBC did a feature program that was videotaped by Nam June Paik. By manipulating the pre-recorded images (including the sound) with magnets, he created a portrait that equals a critical and ironic cartoon. On the whole, McLuhan's influence on the interested art community was diffuse. It appears to have been restricted to a few of his ideas, but the subject deserves further research. McLuhan was apparently a fascinating person whose charisma effectively radiated when lecturing in public. Although his fame and fortune waned quickly during the seventies, he deserves merit for being the first author who created a conceptual framework for understanding the new electronic media. He was also among the first who recognized the impact that these media might have on our perceptual senses, as well as the implications this could have for the education system at large. Paik's visions were congruent with McLuhan's arguments in that the electric or electronic technology (McLuhan used both terms indiscriminately) and its consecutive automation techniques and computerization would bring about a major change in the concept of learning and knowing. McLuhan's non-hierarchical treatment of the different art forms may also have appealed to Paik at the time, being in tune with the artist's own approach.[303]

Video as Communication

Paik has put his ideas in writing since 1963, although he was most prolific between 1965 and 1974.[304] During these years, the artist published his visionary ideas concerning television and other electronic media as *the* communication and information technologies to dominate and determine our daily environment. In subsequent years, Paik kept publishing extensively about television's role as *the* new information structure of our society and the possible function that art could play, although his videotapes expressed his concerns even better, as they were always created with television in mind. His writings reveal a great belief in the potential of these new media, whose possibilities are explored by him on different levels. In the first place, they would open new options for the artist as producer in areas such as television, video centers, and network systems.[305] He believed in a vast virgin land lying before us with ample space to experiment. That artists would collaborate on this development interdisciplinary with engineers and technicians was a matter of course. Secondly, he held the opinion that video and the computer would play an influential role in the field of education. Paik: "Combinations of computers and beautiful color TV synthesizers will be an effective teaching machine for computers, media, TV, art and man-machine relationships in general."[306] Thirdly, he was convinced that the electronic audio-visual media, being the feature of American mass culture, would revitalize so-called "Europe-imported High Art" through their combination of abstract, surreal and realistic images, creating a new aesthetics. This "Videoland" of communications media on the rise would enable the growth of a video culture which would, if modeled after the European Common Market, argued Paik, "strip the hieratic monism of TV culture and promote the free flow of video information through an inexpensive barter system or convenient free market." This utopian belief in the "free flow of video information" was put forth in an interesting article titled "Global Groove and Video Common Market," in which Paik elucidates the effects that a "Video Common Market" could have on world peace and the survival of the "Spaceship Earth" (Fuller), the ecology, education, and the progress of knowledge. He continued that "ecology is not 'politics' but a devoutful Weltanschauung, which believes in world design, global recyclization, the shift of our attention from 'you OR me' to 'you AND me', as Mr. Fuller, the guru of the whole movement, never ceases to emphasize."[307] Paik insisted that a major factor in the world's economic and political problems lay essentially in the failure of communication, leading to wars like World War II or the Vietnam War. With McLuhan and Buckminster Fuller, he was convinced that one had to think globally now. For Nam June Paik media information was 'free information.' In accord with this, he uses pre-recorded television programs, videotapes and films by friends and colleagues, and has continuously recycled his own material. After all, talking about the 'uniqeness' of a work of video art was a contradiction in terms to him. So were the concepts of authorship and authenticity, which Paik has consistently flouted.

Nam June Paik has always shown a sincere interest in education, recognizing the necessity to introduce the new media into the education system - these media constituting the most influential communication systems - so that (art) students would familiarize themselves with these technologies and learn to understand their wider social and political implications. He ventilated some ideas about the integration of new technologies in the American (art) education system when he elucidated how television and video could, for example, be used for a "Global University," referring to McLuhan. He proposed ways to use the interdependent electronic media to exchange cultural information universally, philosophizing about possibilities to visualize medieval, classical and electronic music to make them more accessible.[308] It is important to note that he never solely discussed the aesthetic possibilities of the television and digital image, but always tried to create a context to connect the art with the technological, and scientific developments, to create a dialogue between video art and television, between art and politics, or East and West.

A letter to the New School for Social Research in New York, written as early as 1965, indicates the direction that Paik thinks of at the time. For that reason I quote this letter at length: "Projects for Electronic Television, I hope to open a studio for electronic color television in New York City so that I can begin more complicated technical experiments such as maximum exploitation of shadow mask color TV picture tube, self programming of whole video signal through TV cameras, tape-recorders (visual and audio), the combination of electronic music and electronic TV, and if possible, combining the TV with computers and self-invented 50 channel data recorders. As an adjunct to these experiments I plan to construct a compact version of electronic TV for concerts so that it can easily be transported and demonstrated in colleges. It will have unprecedented education effects since it bridges two cultures, appealing both to artistically and scientifically minded people. These two projects of experimentation and education are aimed at a third stage, development of an adapter with dozens of possibilities which anyone could use in his own home, using his increased leisure to transform his TV set from a passive pastime to active creation."[309]

Many of the artist's visions must have seemed rather utopian at the time of writing. Already in 1974 he said concerning *Media Planning for the Postindustrial Society*: "the building of new ELECTRONIC SUPERHIGHWAYS will be an even bigger enterprise. Suppose we connect New York and Los Angeles with a multi-layer of broadband communication networks, such as domestic satellites, wave guides, bunches of co-axial cables, and later, the fiber-optics laser beam. The expenses would be as high as moon landing, but the ripple effect 'harvest' of byproducts would be more numerous."[310] At any rate, Paik's presentation at the 1993 Biennale di Venezia was not without irony. Entitled *Electronic Superhighways: Bill Clinton stole my idea*, the artist comments that in 1974 he had proposed the idea to the Rockefeller Foundation, made $12,000 and subsequently had it printed in Germany, where 3000 copies were sold. "Maybe Bill Clinton read it in the Oxford Library. He used exactly the same terminology."[311]

Although we are still far away from these positive environmental returns today, global communication networks are spreading rapidly. More and more universities and colleges are surfing the Internet. In the world of finance and entertainment, the whole system of communication will soon totally depend on integrated digital computer network systems. However, if the word utopian is used as the German philosopher Karl Mannheim defined it, i.e. as a vision that can be realized under specific circumstances rather than as a vision which lacks any chance of ever coming into being, then Nam June Paik's Utopia falls in the latter category.[312]

The Paik-Abe Synthesizer and Other Collaborations

When Nam June Paik returned to Japan for a year in 1963, he met two engineers, Hideo Uchida (president of UCHIDA Radio Research Institute, a genial avantgarde electronician, who discovered the principle of the transistor 2 years earlier than the Americans, wrote Paik) and Shuya Abe, who would be his collaborators on a number of future projects. The first result of their collaboration, the *Electronic TV & Color TV* experiments (1963-1964), was presented at the New School for Social Research in New York in 1965. It was followed by a series of research experiments to connect the electronic television with the video tape recorder, and they managed to do a trial preview of Paik's electronic video recorder at the Café Au Go Go, on Bleecker Street, October 4 & 11, 1965.[313] In 1964 he had met cellist Charlotte Moorman with whom he would work for more than ten years. She did a few topless performances dressed only in two small video monitors, which did not go unnoticed.

The interest and support of WBGH-TV in Boston enabled Nam June Paik to work on a video synthesizer, together with Shuya Abe. The audio synthesizer had already existed since 1965, but the two sought to combine synthesized sound with synthesized images. Paik was not the only one interested in this area. Stephen Beck and Eric Siegel had been building similar machines that were meant to create synthetic images. However, Paik's goal was to construct a machine that would synthesize and manipulate existing images. In the beginning of 1970, Paik and Abe presented the Paik-Abe Synthesizer for WGBH-TV in Boston. A live-broadcast presentation followed in the summer, *Video Commune*, in which Shuya Abe and other guests 'played' the synthesizer under Paik's direction. The sound of the program was prerecorded and contained the whole oeuvre of the Beatles. Paik considered the video synthesizer as a medium, just like any other material or apparatus, and 'gave' the discovery to other studios as well as art institutions and colleges, which were soon able to integrate the synthesizer in their program or curriculum. This gesture is again indicative for Paik's point of view on ownership of ideas or originality. Paik later: "I was very proud at the Media-Art Conference in Washington D.C. on December 2, 1972, because the Paik-Abe video synthesizer was played not by Mr. Paik but by Mr. Ron Hays. If a machine is to survive as hardware, it should be universally applicable...like an automobile, which anyone can drive. So far the Paik-Abe synthesizer is the only video synthesizer being used cross-country (WGBH, WNET, Binghamton TV Center, Chicago Institute of Arts, California Institute of Arts at Los Angeles) by more than 100 artists and it has been aired locally and nationally quite a few times."[314]

Around 1968, Nam June Paik had made contact with computer technicians at Bell Telephone Laboratories, Murray Hill, NJ, to do computer research as a "residential visitor" under the guidance of Michael Noll, a programmer-engineer who made some important contributions to the developments in computer graphics and also wrote about computer art related issues. At Bell Telephone Laboratories he also met engineer Bob Diamond, and their collaborative efforts led to the connection of the digital computer to the video synthesizer.[315] Expressing his ideas about the collaboration between artist and engineer in an interview with Jud Yalkut, he said: "I also envisage the day when the collaboration of artist and engineer will progress into the unification of artist and engineer into one person. According to my past experience, the best results were achieved through accident and error. As you see, the transistor was discovered by accident. This means that the present computer age was the product of chance to a high degree. Therefore, if I give an order to an engineer, and if I don't go through all the experiments myself (that is, the complicated process of trial and error), I will lose all these precious errors, I will only get what I want, and miss all the disappointment and surprises. I have found that the by-product is often more valuable than the first envisioned aim."[316] This dialogue took place in 1968.

Throughout his career, Nam June Paik has collaborated with engineers, technicians, producers, and artists and friends. And not only has he always acknowledged the fact that the production of videotapes and multi-monitor installations is a collaborative process, he is even convinced that collaborations result in more imaginative discoveries and 'masterpieces' than brooding alone in the studio. Thus he tells how his first collaborator, a "teenager named Guenther Schmitz" whom he met in Cologne, "guided me through the discovery of horizontal and vertical modulation and other techniques which became the scan-part of the video synthesizer later on." Playing down his technical knowledge, he maintains that he never even played the Paik-Abe synthesizer himself. Freely acknowledging his ignorance (is he really that ignorant?) about the workings of a televison studio he continues "In the making of 'The Medium'... I was in complete panic. So I told Fred Barzyk and David Atwood: 'I am not here, do whatever you want. In any case I don't know anything about it.'" Since 1982 Paik has collaborated with Paul Garrin. In the same article he continues: "Anyway my collaboration with Paul Garrin is like an improvisation of a 4man jazz ensemble. ... If the first tenor is a new machine, the first soprano is Paul Garrin, a guy who is at least 200 times faster than me in those machines. That means in the first $1,000 dollars-a-hour machine time expense he can produce 200 times more than me."[317] Everybody Paik works with is always fully credited.

The distribution of the Abe-Paik synthesizer to other television stations and colleges shortly after its invention also tells something about Nam June Paik's attitude toward ownership of ideas, of images, and of inventions. From the start, Nam June Paik utilized existing imagery, original television footage, films, taped lectures and concerts, anything, in combination with material he had taped himself. He has continuously recycled his videotapes. In the same way, he keeps revising his installations, creating multiple versions. For example, his most renowned works *TV Buddha* (1974) and *TV Garden* (1974) exist in a number of variations. Yet their main theme remains the same. The original *TV Buddha* shows the seated Buddha reflecting his own mirror image on the monitor in front of him. *TV Buddha* is a closed-circuit installation, with a camera recording the statue whose image is fed back into the monitor. For *TV Buddha No. X* (Whitney-Buddha-Complex, 1974-1982), Paik covered the monitor with an earthen pyramid, with only the monitor screen left free. The Buddha statue used for this version was a Korean one from the 7th century. In *Stone Buddha* (1974-1982), the mountain consisted of natural stones, while Buddha no longer watched himself, but *A Tribute to John Cage*, an earlier videotape. For *TV Garden* (1974), Paik arranged a 'garden' of tropical plants, among which he placed 20-30 monitors. In the first version, he planted some one hundred tropical palm trees in between 28 color television sets. The television sets were laid down randomly, screens upwards. The images were not completely visible among the plants, which created the impression of "television sets (that) function as organic forms." As *TV Garden* imagery, the artist consistently showed his *Global Groove* videotape.[318] By continuously changing the same or similar imagery, Paik shows how vulnerable our visual impression is to making us believe that we perceive something new all the time. Paik's videotapes are a reflection of how in effect we constantly re-arrange the signals received into messages and into new information,. Television and video are the media that technically represent this quality. The images can be re-edited, put into another sequence, or deleted; they are temporary, indefinite. Nam June Paik has consistently pointed at this basic quality of video.

Participatory Television

Nam June Paik's research has always connected a theoretical position with a practical (technical) solution, and his solutions have always consisted of a layering of concepts that relate the internal structure of the work, whether videotape or installation, to the outside environment, be it a television or a satellite program, or a presentation in a museum or gallery. For an artist involved with communication and control theory, the structure of television as a one-way information stream had, ideally, to be changed into a two-way communication system.[319] His first steps involving the public actively started with the *Exposition of Music - Electronic Television* at Galerie Parnass, where the public was invited to 'play' the *Klavier Integral* or the *Urmusik*, and create their own television program as it were on some of the television sets exhibited. The participatory element had quickly become fashionable, in the late fifties and sixties, when artists tried to engage the public in happenings, performances and subsequently environmental type installations. Paik defined it "as the next step toward more indeterminacy, I wanted to let the audience act and play itself."[320] Actually, what the artist is saying here is that the audience was left free to create a work of its own. This concept again places Paik ahead of his time, for it presages his views concerning the authenticity of art and authorship. It is hard for us to imagine its potential, for the few remaining objects from that period that have found their way into museums cannot be touched any more.[321] A more elaborate audience participation was developed a few years later when he experimented with placing magnets on television sets such as *Magnet TV*, and *Participation TV I* and *II* (1965 and 1969), creating distortions in the electronic signal. In the case of *Magnet TV*, the artist had placed a magnet on the screen with which the viewer could change the transmitted image electromagnetically. It was developed more or less accidentally, for, while Paik had experimented with magnets inside the electronic circuit, it had not occurred to him to apply it simply to the screen.[322] *Participation TV II*, created for the first exhibition of video art, *TV as a Creative Medium*, at the Howard Wise Gallery (New York, 1969), was a closed-circuit installation of three cameras connected to the red, blue or green cathode ray tubes. The images of the color-separated cameras created a three-color layering on the television screen, multiplying the image as ghost-shadows, like an echo. If the camera was directed at the monitor, it faceted and fragmented the closed image circuit in a total feedback loop. [323] Although these works involved the viewer more actively, one cannot yet truly speak of a viewer participation in the sense that he or she was able to transmit a message of his or her own. The same was true for another famous closed-circuit installation, Frank Gillette and Ira Schneider's *Wipe Cycle*, which was also shown during this exhibition. Both remain landmarks in the history of video art, however, as they coupled the inherent technical qualities of this new medium with an extended context, that of the viewer.

Viewer participation has remained one of Paik's continuing concerns, although it often remained restricted to making the viewer aware of the viewing conditions by setting up situations which required the visitor to become physically involved. For example, Paik had his spectators lie on a *TV Bed* (1972), or on their backs watching *Fish Flies on Sky* (1975). Paik felt that people could create their own art by sending video to their friends by video-telephone lines to "elevate their mood by watching or attaching certain medical electronic gadgets and control their own brainwaves in order to achieve an instant Nirvana." These ideas are congruent with the artist's idea that we should get rid of the hierarchy between 'high' and 'low' art, that it was time for the distinctions between creator and audience to disappear. Meanwhile, he has created works in which he uses the principle of interactivity, made possible by new interactive computer technologies (San Diego).

Overkill Strategies

Global Groove (1973), a classic not only among Paik's videotapes, but also in video art history was broadcast on January 30, 1974, and opens with a prophetical sentence: "This is a glimpse of a new world when you will be able to switch on every TV channel in the world and TV guides will be as thick as the Manhattan telephone book." In *Global Groove* show-business elements, disco and go-go dancers, Mitch Ryder and the Detroit Wheels' Devil with a Blue Dress On are alternated with Korean drum music, American-Indian chanting, Allen Ginsberg and John Cage telling anecdotes, films by Robert Breer and Jud Yalkut, and segments of earlier performances (*26'1.1499" For a String Player)* and videotapes (*McLuhan Caged*, and *Electronic Opera No. 1*) This videotape became a model for his future works. It deals with the television format (screen), its communication structure, and the viewing habits of the American public on a number of levels. First, there is the fast editing pace that Paik adopted (or adapted to?), which is, however, sometimes a little too fast, and at other times a little too slow.[324] Secondly, by using the single fast-cut image format for a wide range of non-related subjects, he created a high-density level information structure. Since there is no recognizable narrative, the viewer only has time to be engulfed in the overload of imagery, without being able to attach a specific meaning to it. In *Global Groove,* representational and real time imagery, images edited in a fast-paced non-sequential rhythmic structure and synthesized, colored bits alternate and correspond to three levels of information, which are described by Paik as the work/life state, the dream state, and the heightened consciousness state. It shows the artist's innate understanding of the language of television, in particular the dimensions of its time structure, so that he can subtly manipulate its formulas, and even 'control' the perceptual qualities of the medium as an image-making device.

While Paik utilizes and accepts the entertainment aspect of television with its information overkill, he shows himself critical of the system at the same time. In adopting the same formula as television does, but with subtle modifications, he plays with the medium's perceptual qualities on the one hand, by consciously applying his own formal structures as methods of manipulation and exploitation, the global communication medium that television wants to be is questioned on the other hand. David Ross once interpreted this ambiguity as: "It becomes evident that Paik relishes the contradictions inherent in the very interface of an uncompromised aesthetic and politic within a context of total compromise that IS broadcast TV."[325] After all, it was at WGBH-TV Boston that Paik produced his first videotapes, *The Selling of New York* (1972) and *Global Groove* (1973).[326] The collaboration with WGBH-TV Boston was discontinued in 1972, after a program called *Video Variations*, produced by Fred Barzyk, with artists James Seawright, Stan VanderBeek, Russel Conner, Douglas Davis, Mimi Garrard, Constantine Manos, Jackie Cassen. Paik's contribution was entitled *Electronic Opera No. 2* (1970). The collaboration was renewed with *A Tribute to John Cage*, made on the occasion of Cage's sixtieth birthday in 1972. In this tape, John Cage presents and discusses his ideas and work. The composer specifically re-orchestrated *4'33" (Tacet, Tacet, Tacet)*, originally performed by David Tudor on August 29, 1952, and now played by Cage himself on Harvard Square at Woodstock. The musical sequences alternate with excerpts from previous videotapes, such as *Chroma Key Bra, Video Commune, The Selling of New York*, and *Variation No. 5*, and interviews with friends. *A Tribute to John Cage* shows the typical Paikian way of filming and editing; a fast pace with an overkill of images is alternated with slow fragments. The imagery covers a broad cross-cultural spectrum. Here as elsewhere, dance functions as a metaphor to bridge different cultures and make international communication possible.

Working at WGBH-TV allowed Paik to closely research the television structure with which and against which he has worked since. His investigation of electronic imagery kept revolving around its meaning for our culture industry, an industry permeated by and totally "dependent upon the electronic media for communication, instruction and pleasure." Paik said to Douglas Davis in 1973: "But in the future world, you will have cable TV, video cassettes, and picture phones. You will also have video tranquilizers, like my Synthesizer Machine. If the video structure is diversified, one of the first results will be less pollution. It will be a major solution to the ecology crisis. Why move, why drive somewhere in your car, if you can do everything right at home."[327]

Paik's overkill strategy expressed itself in video installations which became larger and larger, with image frequencies so fast that they appear to flicker before the eye. Initially *TV Garden*, with its 20 to 30 monitors, looked impressive. *Fish Flies on Sky* first presented in

1975, already used 88 monitors and 2 videotapes. The multi-monitor installations of the eighties became gigantic by comparison. Among the most majestic pieces was *Tricolor Video* (1982), conceived for the Centre Pompidou in Paris: 384 monitors were laid out in a square on the floor, in twelve rows of eight. These rows were divided in three, one color from the French flag dominating in each section of four rows; blue, white and red respectively. To complicate matters, Paik had designed an image program with 8 videotapes running diagonally. The videotapes contained sections and images from *Global Groove*, *Suite 212*, *Guadalcanal Requiem*, etc. As usual, there was material from television programs and films, among others *Casablanca*, and even sequences from videotapes by other artists.[328] Often the work was adapted to the architectural space, like *TV Trichter* (Düsseldorf, 1984), where 99 monitors spiraled upwards in a cone shape, screens downwards.

The peak of this development is certainly formed by the enormous video wall conceived for the Chase Metrotech Lobby in Brooklyn, called *Chase Information Wall*, in 1992, and *Sistine Chapel*, first shown at Holly Solomon gallery in New York, 1993. The *Wall* includes computers, laser discs, 429 television sets in different formats, and cable TV. The *Chapel* is a multi-channel projection piece. Video projectors and speakers were placed randomly on a large scaffolding in the middle of the room, beaming images onto the walls and ceiling, like in the real Sistine Chapel, in a fast rhythmic staccato pace. The sensory overload reminded one of rock concerts, or the now popular techno-parties. Still one could discern the figures of Paik's long-time friends who had died recently: Joseph Beuys, John Cage, Charlotte Moorman and Alvin Ailey, interspersed with landscapes, the American flag, sports events, and schools of swimming fish.

Opposites with Crossovers

Nam June Paik's work appears to be dominated by opposites. Opposing his mega-installations are the closed-circuit pieces like the buddhas, the egg and fish works. In these installations a single motive dominates, in which both the formal and the symbolic or metaphoric qualities define its meaning. Also, these motives keep recurring. *Positive Egg* and *Negative Egg* (Newport Harbor Art Museum, Newport, 1993) formally go back twenty years, to installations like *Moon is the Oldest TV* (1976) and *TV Clock* (1977). In *Positive Egg*, the camera is directed at a white egg which lies on a black cloth. On a series of monitors which become successively larger, the form of the egg grows in size until it seems almost an abstract oval. Similarly in *Negative Egg*, a boiled egg cut in half becomes larger and larger. Appearing simple and meditative, these minimalist installations have grown in beauty and elegance. Maybe it is also a recognition of video's essential quality, that of monitoring. There is a similar reductionist tendency in the recent works of Bill Viola and Paul Ryan. A sharper contrast to the nineties 'baroque' multi-monitor installations, such as *Buddhamorphosis* and *Paranature* (both 1992), is hardly imaginable, extravagant and overly decorative, as those are. Critic Kenneth Silver commented: "This gaudy agglomeration of elements is united by the shared theme of stage and spectatorship: the Buddhist shrine (with its hinged doors and cult image within), the puppet theater and televison are posited as three related sites."[329]

As said repeatedly, certain images keep evolving over the years, like the Buddha, the egg, moon and fish. Despite their formal qualities, Paik has obviously also their symbolic aspects in mind - as related to the medium video. Being characterized by constant reconceptualization and manipulation of formal elements, Paik's work appears to be based on a formula of opposites with crossovers. Appears, because they are never true opposites, nor irreconcilable. Rather, they are complementary, like the yin and yang symbol; herein one may recognize the Asian and Buddhist in Paik. The elegiac contrasts with the overload strategy. The single motive closed circuit installation opposes the rapid fast cut non-narrative sequences of multiple images, the sacred meets the profane, nature features in technology, East infiltrates West.

As Paik pushed his investigation into electronic television further, his field of research expanded to comprise a range of concerns, including the social and political function of television, the role of the viewer and viewer participation, and the function of the electronic and computer technologies as communications media and global networking systems. To set up live television programming, simultaneously broadcast via satellite in different continents was one of his early visions. In fact, already in 1966, he designed a *Utopian Laser TV Station*, which was published as a manifesto, in which he envisioned that "Very, very high frequency oscillation of laser will enable us to afford thousands of large and small TV stations. This will free us from the monopoly of a few commercial TV channels." The idea of a global communication network certainly sounded utopian, then. But with the technology already being developed at the time, the possibility existed in principle. The first satellite experiments had taken place in 1960 via Echo 2 and Telstar 2, and television satellite broadcasts between the United States and Europe started in July 1962. Things developed slightly differently from what Paik thought, and they have not brought active viewer participation much closer, except in such rare instances as the experiments of "Van Gogh TV," presented by the Ponton group in Germany. But the world is covered with satellite dishes now, which indeed make multiple channel reception possible world-wide. In BBC's *Saturday Night Clive*, the viewer travels with Clive James in real-satellite-time in thirty minutes from London to Hollywood/Los Angeles, to Sidney, to Manilla in the Philippines and back in an hour. Via satellite the latest news on the Gulf War was brought into the living room without delay, and we saw CNN's Peter Arnett report from the roof of the Rashid hotel in Baghdad. The Gulf War was to a large extent a media war as well. Not only has this medium changed our view of war, and the world, through a direct, simultaneous stream of news information. It had a direct impact on the way news is presented, just because of its immediacy.

It should not come as a surprise that the artist has also ventured into this realm. Although Paik had investigated the possibility radiocasting a piano concert simultaneously in San Francisco and Shanghai in 1961/1962, not before *Documenta 6*, Kassel, 1977, was Paik able to set up his first satellite program. The next satellite project was *Good Morning Mr Orwell*, on 1 January, 1984. As locations from which the broadcast took place were chosen the Centre Pompidou in Paris and a WNET-TV studio in New York City.[330] The program consisted of live performances by (avant-garde) artists and pop-rock musicians. The structure of the program was comparable to that in his videotapes: a fast-paced collage of heterogeneous real-time, electronic and synthesized images. Among the participants in Paris were Ben Vautier, Joseph Beuys, Robert Combas, Yves Montand, and the group Urban Sax. In the New York studio were Laurie Anderson, John Cage, Merce Cunningham, Peter Gabriel, Allen Ginsberg, Philip Glass, The Thompson Twins, Mauricio Kagel, and Charlotte Moorman. Paik used the split-screen technique to show what was happening on both sides of the Atlantic simultaneously. Criticism of technical failures, of delays and inconsistencies abounded. However, in being intercontinental, in combining elements of entertainment and contemporary (high) art, interjecting multi-cultural subject-matter, and as an interdisciplinary collaboration, *Good Morning Mr. Orwell* was for Nam June Paik a model for satellite television as a social, political and educational communication system, in which he embodied his ideas about television and video as communication media that might truly contribute to a better communication among humankind, and lead to a better understanding of the world's problems.

The theme of communication runs like a red thread through Paik's videotapes and video installations, for, in his view, war and global conflicts are primarily the result of a lack of communication. This is also why his works contains a layering of multi-cultural elements (dance, poetry), ideas and symbols (moon, egg, fish), and historical and political features. The elements of nature and state-of-the-art technology are of equal importance, both being a part of our daily lives.[331] Nam June Paik will broadcast his last TV show live on January 31, 1999.

Chapter 11

Bill Viola: A Close Re-View of Reality

Beyond the Frame of the Screen

Bill Viola belongs to the first generation of video artists who have devoted their artistic career solely to the medium of video. Viola was born in Flushing, New York, in 1951. His early involvement with contemporary music enabled him to meet many composers of "new music," like David Tudor, with whom he studied and performed between 1973 and 1980. The artist's background in electronic music would keep playing a role in the video works.

Viola has said that he began to experiment in these media because he wanted to get out of the studio. He happened to be at the right time at the right place: Syracuse University, where Viola studied, enjoyed an active video environment at the time. The Everson Museum of Art was just organizing its first video exhibitions, which culminated in the Nam June Paik retrospective in 1973. Viola himself became involved in setting up the university's first cable TV network, where he had his training as a studio engineer and became a member of the Synapse Video Group. Upon his return from Florence, Italy, where he worked as technical director at Art/Tapes/22, one of the first experimental video studios in Europe, from 1974 to 1976, he obtained the first artist-in-residency at WNET-TV's Television Laboratory / Channel 13, in New York (1976-1983). This enabled him to explore and work with the sophisticated computer editing system there. His early works in video, like those of most video artists, basically were an exploration of the technical features of the medium, such as real time, feedback, closed-circuit, as well as synthesized imagery. The importance of video for him was its possibility to view at once what one had recorded a few seconds before, making an instantaneous reaction possible to the experience just before.

Viola also claims to have been influenced more by experimental filmmakers of the time - such as Stan Brakhage, Michael Snow, Jonas Mekas - than by the contemporary videomakers. The work of experimental filmmakers, like Michael Snow, focused, among other things, on how structural devices could influence or even determine perceptual qualities. And like Brakhage, Viola rarely used spoken language, and the few times when this happened the words were "just ... part of the sound landscape."[332] Nam June Paik had expressed himself quite simply - "film is chemicals, TV is electronics. There are something like four million phosphor dots on a twenty-one-inch television screen every second; it is just like Seurat - you mix them in your eye. In film, you take from reality; in TV, you produce reality - real electronic color." But Bill Viola's early concerns were truly in trying to define technical and aesthetic qualities inherent in the medium. Just as the experimental filmmakers were exploring the material film some decade earlier, Viola began experimenting with the materiality of video, investigating the nature of the electronic moving image, and especially the component of time.[333] To the French theorist/critic Raymond Bellour he said: "The crucial thing for me was the process of going through an electronic system, working with these standard kind of circuits which became a perfect introduction to a general electronic theory. It gave me a sense that the electronic signal was a material that could be worked with. This was another really important realization. ... I never thought about (video) in terms of images so much as an electronic process, a signal."[334]

He never cared much for the idea of a 'guerrilla television' which saw its heyday in the early seventies, with numerous community-oriented groups like Videofreex, Global Village, Raindance Corporation, etc. searching to influence the commercial network structure with alternative television. This did not mean that he did not deal with television. However, in contrast to Paik's overkill strategy, Viola preferred to deviate strongly from the formal structures and editing concepts that have become television's mode and next to the former one's videotapes, the latter one's even look sparse, with their often slow-paced editing sequences, seemingly devoid of technical effects. Seemingly, because - like Paik - from the start Viola has sought to use state-of-the-art technology. After he was offered an artist-in-residency at the Television Laboratory of WNET-TV, Viola has always produced broadcast quality videotapes, having in mind that they would be broad- or cablecast. And most of them were. Television is the country's most powerful news medium, and extremely influential. For Viola's generation, television functions as society's conscious reflection and moral self-image, and he has always been convinced that artists should be using this medium, because video, "when coupled with the human mind can offer us sight beyond the range of our everyday consciousness, but only if it is our desire, both as viewers and as creators to want to do so."[335]

Viola was also concerned about the 'narrow' territory of the visual arts and looked at what was going on in other fields, including psychology, anthropology, cybernetics, philosophy, history and so on, to connect video's internal communication aspects with a much broader interdisciplinary field of communication. And like other media-conscious artists, he was well-informed about the intellectual ideas of Marshall McLuhan, R. Buckminster Fuller or Norbert Wiener. Viola: "To see art as part of this larger context is to realize it is only a small part of a larger picture. The first time I went out there was very important in that it broke down the boundaries of thoughts that held me within the art world itself. It's like a figure/ground reversal. I saw the art world as this little category that we've made a space for in the larger cultural structures we've created."[336] Being involved with television's modes of communication constituted one possibility transcending the "boundaries of the art world" - as a medium that, both technically and structurally, asked for a viewpoint that would take into account the broader social, political, and cultural implications; as a means to be used as a cultural strategy. Subsequently, Viola began to ask himself questions about other qualities of the moving image: the nature of human perception and our capacity to memorize things. How do we retain images, for example? Why do we remember certain occurrences in our lives, and forget others? How does this influence our understanding of history, and time? His wide-ranging studies in the scientific topics of the day, in 'foreign' cultures, in the literary and philosophical history of the West and the East to find cross-cultural relationships, served to acquire the background knowledge for a more specific research into the nature of memory and perception. Bill Viola's interest in memory, and time, have led him to research the functioning of the brain, i.e. the functioning of our mental faculties. He familiarized himself with the developments in artifical intelligence research as well as the latest ideas in neurophysiology and neural networks.

Since the sixties there has been a continuous research in man - machine (i.e. computer) relationships, and investigations into the possibilities of providing computers with learning capabilities. Whether the computer will or even can take on the same function as the human brain is not that important here. What concerns Viola is that the digital and electronic technologies are representative of the current scientific way of thinking. This is a major reason for him to use video as a tool, for he considers it representative of the state-of-the-art 'system of logic' or 'philosophy.' Viola: "Consciously or unconsciously, most people assume the existence of some sort of space when discussing mental functioning. ... This mental space is directly analogous to the "data space" in our first brainchild, the computer, being the field in which calculations occur and where virtual objects of digital graphics are created, manipulated and destroyed. Like a fundamental ontology, this given space is perceptually before or after what is done, an a priori existence from birth in the flip of a switch until the lights finally go out."[337]

Time-Space Continuum

While his early works evolved from an investigation into video techniques, a major change occurred after his travels to the Solomon Islands in 1976. Having set out on a UNESCO-funded project to train native islanders in the use of video recording, he also documented some of the traditional music, song and dance of Melanesian tribes, resulting in two videotapes: *Palm Trees on the Moon* and *Memories of Ancestral Power: The Moro Movement in the Solomon Islands* (1987). The latter tape is a seemingly straightforward anthropological documentation. Yet the emphasis is on images as prelinguistic carriers of symbols, myths, and memory, as a means to preserve tradition, and so prefigures his now continuing concern with cross-cultural significance. The memory of the past and its continuation into the future as oral history are the expression of a concept of time which is determined by the human life cycle. Since our western history has come to depend almost completely on the written tradition which predominantly functions as a chronological, linear concept of history, and thus time, the possibility that traditional knowledge can be handed down orally belongs to a gone-by era. This realization led to numerous questions and research into other (than western) cultures; about what these could mean to us, and what we could learn from them.

It goes without saying that the element of time in video as an inherent technical quality was of great significance as a formal device, which was yet enhanced by the introduction of the computer, enabling time-code video-frame editing (ca. 1973/4). The new digital editing facilities, to which Viola had access since 1976, replaced the real-time, linear representational composition and made it possible to compose the images in any time structure desired. For example, in *The Reflecting Pool* the frame is broken up into three distinct levels of time: real-time, still, and time-lapse. Time is constructed to look like a complete image of a single space. Viola refers to it as sculpting with time. *The Reflecting Pool* are five independent works that function like a cycle, evoking birth, growth, death, rebirth. They are: *The Reflecting Pool, Moonblood, Silent Life, Ancient of Days*, and *Vegetable Memory*. Internally, the works also contain cyclic or rhythmic layerings, such as the human life cycle, natural cycles, the daily cycle, moving from day to night to day. Thus *The Reflecting Pool* (1977-1979), deals with the fact that all life on earth is subject to the same processes, whether perceived as linear, cyclic, or concentric. In *Ancient of Days* (1979-1981) time as duration is the major structuring device. Using effects like slow motion, reverse motion and time-lapse, Viola appears to want to make time a tactile experience. *Ancient of Days* (1979-1981) was described by Viola: "Diverse rhythms of natural and subjective time are interwoven into a complex whole using a mathematical notation of time code editing. Notions of temporal symmetry and duration transposing are explored, along with extensions of video time-lapse techniques. ... The piece is the result of research conducted on the computer editing system at the Sony Corporation, in Japan."[338]

For Viola, investigation into natural phenomena, i.e. the levels of existence as being mineral, vegetable, plant, animal, man, etc. are always related to how we as human beings perceive them. Few people realize that we experience something differently each time we view something, so that in fact we partially recreate the mental image of our own environment continually, both physically and mentally. *Vegetable Memory,* which title is derived from the writings of Jalaludin Rumi, a Persian poet of the 13th century who moved Viola deeply, explores the perceptual effect of repetitive cyclic viewing. A cycle of images - recordings from the Tsukiji fish market in Tokyo - is repeated at different speeds, so that the images appear extended in time, and because of that appear changed in form, more pictorial. Images of fish, dying, cut, dead, in buckets, frozen, handled by hands of fishermen, do not just remain images. This non-narrative technique affects the viewer's perception in that it offers options for personal associations and subjective meaning apart from the known symbols and metaphors they refer to, bringing back memories and creating new ones. In addition, the place being the fishmarket, its surroundings and the activities there make us part of an expanded context, the reality of a social and economic situation. Viola: "The idea of vegetable memory is going from one level down to the next one, in the sense that there are these images of fish which have been taken out of where they are and what they are (the water), and brought to the next level (the land), which for them is death. Land is death for the fish in the same sense that for us there is another world above the surface. ... It's the idea of death in the positive sense, in a sense of growth and not in a negative sense." It is a recurrent metaphor in the artist's videotapes.[339] Finally, said Viola, the series is also a metaphor for a personal journey of the artist. The works, as it were, attempt to connect the manifold layers of time as they unfold in the different time sequences and the workings of electronic videotape, to understand time cyclically by viewing (playing and rewinding the videotape). The viewing experience is after all also a mode which is literally cyclical.

Levels of Perception

By that time Viola had reached a point where he realized that anything he did with the medium had to reach another person via the channels of perception, and that this image system which is part of our bodies, so to speak, was just as important as the hardware itself. Viola's images of nature are not only a representation of nature as it really is. He continually investigates and questions modes of perception: how do we perceive, what are the limits of our perception? His body of video work reveals this continuous search to try and find the connecting forces between nature as it is perceived and understood, an investigation into "the mind's eye," as he calls it. *Chott el-Djerid (A Portrait in Light and Heat)* (1979) "confronts the final barrier of the limits of the image. At what point does the breakdown of normal conditions, or lack of adequate visual information, cause us to re-evaluate our perceptions of reality and realize that we are looking at something out of the ordinary - a transformation of the physical into the psychological?"[340] *Chott el-Djerid* opens with an long shot of a winter snow landscape in Saskatchewan. The blue color of the sky is set against the white of the snow. As there is no depth of vision, the image is barely distinct from a flat piece of paper painted blue and white. It is only the dark shimmer on the horizon, or the snowstorm, which makes the image visible as landscape. From Canada, we switch to the Chott el-Djerid, a dry salt lake in the Tunesian desert, known for its mirages. The heat creates conditions that distort our vision, disorient, build uncertainty about what we really see. The images seem unreal, hover between illusion and reality, between appearance and disappearance. The natural elements change into each other and become similar. The branches of a tree resemble a web of nerve cells. Viola: "The reality of the landscape is also an inner reality, literallly and symbolically. ... Underneath was the recognition that electronic space can coincide with real space under certain conditions. That the electronic image could be interpreted as a white snow field with a blue sky."[341] His ideal is to capture something of the essential energy of the place, often waiting long hours for the right atmospheric image. Viola was able to capture these natural phenomena in the landscape with a special telephoto lens. Using one video camera set on a tripod, he recorded from a fixed and meticulously calculated position, occasionally readjusting this position somewhat, so that he would be able to recreate a continuous zoom while editing.

Although it is possible to create these effects artificially with the current state-of-the-art technology, Viola has emphasized that it was important to him that the viewer realize that these images are natural or 'real,' yet that they force the viewer to raise questions as to how we decide what is real and what could be a hallucination, and ultimately what is really the nature of perception. When he studied experimental psychology, for example, to learn about the workings of the central nervous system, and about how the eye works, he found that the whole human system is geared toward the perception of change, i.e. that we see because of changes happening and that this is also why we perceive time. The sounds in *Chott el-Djerid* are natural ambient sounds from the site itself. His background in music has been of major influence on the way he uses sound in his visual works. Thus he compares the characteristics of sound and acoustics, and in particular electronic sound with the electronic image, positing the idea of a sound field, which is defined as a field that is always present. It is a concept that comes from the Hindu philosophical notion which sees the origin of all things in sound, as represented by the essential vibration Om. Being a field there is no central viewpoint. This is also a characteristic of the electronic image.

Viola: “In technology, the current shift from analog’s sequential waves to digital’s recombinant codes further accelerates the diffusion of the point of view. Like the transformation of matter, there is a movement from the solid and liquid states into the gaseous. There is less coherency, but previously solid barriers become porous, and the perspective is that of the whole space, the point of view of the air.”[342] Interestingly, Woody Vasulka, also a video artist of the first hour, said something to that extent: “There is a certain behavior of the electronic image that is unique... It’s liquid, it’s shapeable, it’s clay, it’s an art material. It exists independently. ... The electromagnetic spectrum exists, organized and unorganized totally in space.”[343] I already mentioned before that Robert Smithson created a few sculptural works in which he dealt with elements of obsoleteness of the Renaissance perspective as well. Smithson’s thought evolved from his research into crystalline structures, Viola’s is based on physical properties of electronic technologies.

Both Nam June Paik and Bill Viola understood early on that there was a congruence between the development of the communication and information theories and the ‘evolution’ of electronic technology, after the discovery of electromagnetism by James Clerk Maxwell. Viola has mentioned how the understanding of certain principles, like basic wave theory, amplitude modulation, frequency modulation, carrier wave, how light behaves, etc., were important to him as basic principles from which electronic communications theory could evolve in the first place. The electronic image, television, is based on the same technology (general wave theory), and subsequently made the development of the digital image possible. Thus, while the discoveries in the science of physics had required scientists to perceive time and space as correlates, the new technologies are in fact an application of this new way of thinking. Viola: “The video image is a standing wave pattern of electrical energy, a vibrating system composed of specific frequencies as one would expect to find in any resonating object. As has been described many times, the image we see on the surface of the cathode ray tube is the trace of a single moving focused point of light from a stream of electrons striking the screen from behind, causing the phosphor coated surface to glow. In video, a still image does not exist, in fact at any given moment an image does not exist at all. In terms of energy, the beam is always activated and in motion, and there is always a steady stream of electrical impulses coming from the camera or video recorder driving it. The divisions into lines and frames are solely divisions in time, opening and closing temporal windows that demarcate periods of activity within the steady stream of electrons. Thus, the video image is a living dynamic energy system, a vibration appearing solid only because it exceeds our ability to discern such fine slices of time."[344] In short, the electronically created image has no single point of focus, nor is there any perception of depth possible. The television screen presents an overall flat image. This perceptual synaesthesia, this cross-over and interchangeability between the visual and aural senses is seen by Viola as a natural inclination of the structure of all contemporary electronic media.

I should interject at this point that Bill Viola always uses state-of-the-art technology even if it is not apparent. His continuous research into new developments is just as much a part of his life as is sleeping. Yet he does not perceive his work solely within the context of art, but the broader cultural tradition of anthropology. Marie Luise Syring recently even called his style "*visionary* documentary."[345] For example, *Hatsu Yume (First Dream)* could have been just a straight sequence of images presenting aspects from Japan's natural environment and city life, had it not been for the way in which Viola records his images: the variations in the length of the shots, his mode of timing, the moving up or slowing down the speed of recording, the special way in which he edits his sequences. The tape opens with a sunrise over the ocean and the shoreline, moves to Mt. Fuji, bamboo groves, rice fields, and a hot spring at a mountain shrine, a boulder with small rocks - obviously places with special meaning for the Japanese - to Tokyo's fish market at dawn, a tour through the city at night filmed from a car, and closes with images from fishing boats at night. Without narrative and only using ambient sound, the images receive new meaning, beyond the representational. "First Dream is probably the least literal of all tapes I've done. It's about birth and death, starting with light and ending with light... What we're really talking about is images as symbols," said Viola. "Video treats light like water - it becomes a fluid on the video tube. Water supports the fish like light supports man. Land is the death of the fish - Darkness is the death of man."[346] While in Japan, he and his wife, Kira Perov, lived in a small studio apartment in Tokyo, studying Japanese traditions, like Noh theater, Za-zen meditation techniques, learning calligraphy, even the language, in order to understand the culture, albeit from a western vantage point. *Hatsu Yume* transfers the images that have a specific symbolic meaning in Japanese culture onto a general level, where they provide the Western viewer with some intuitive insight into this culture. Viola chose those images from the natural environment that he thinks still have a symbolic meaning for us as well: mountain, rock, fish, water, light. "The mountain is a traditional symbol for time, eternity, stillness, and I've used it with those meanings in mind, although it's not only that. That rock is time, really: representing rock in slow motion, images moving 16 times normal speed, it slows up to real time and then to slow motion. It's like windows or wavelengths of perception. There are all these different rates of perception that are simultaneous and interwoven at any one moment, but we are tuned only to a certain bandwidth or range. ... It's directly related to scale changes in space or even music."

Hatsu Yume was made during a one year residency in Japan on a Japan-United States Creative Arts Fellowship, which was extended by half a year as artist-in-residence (1980) at Sony Corporation's Atsugi Plant. Sony Corporation allowed him to use their latest high-resolution camera, particularly suitable for low light conditions, a feature that Viola has thoroughly exploited in this videotape. He learned from and worked with the engineers who had designed the new 1 inch computer editing system at the Atsugi Plant. In addition, he developed a special zoom lens motor control with Japanese engineer Yasio Shinohara.

The formal aspects - long shots, alternated with close-ups, the use of slow motion or the opposite - all have strong observational qualities, of course, related to the time-space continuum. Together they are used structurally so that the image becomes a sign for something else. It makes an image or sign transcend into a symbol or metaphor; even gives the suggestion of another kind of reality existing beyond the purely visual. The creation of a time-space continuum is particularly apparent in *Anthem* (1983), a videotape in which the single scream of a child recorded in Union Station, Los Angeles, is elongated in time over the entire tape of eleven and one-half minutes, unifying the diverse kinds of images, related to heavy industry and medicinal technology and daily life habits. In *He Weeps for You* (1976) a drop of water slowly came out of a copper pipe. A color camera with a special lens recorded this process, which was projected on a large screen at a rear wall. In the drop of water the surrounding space was mirrored, including the observer. Time appeared extended. As the artist explained: "The piece makes reference to the traditional philosophy or belief that everything on the higher order of existence reflects, and is contained in, the manifestation and operation of the lower orders. This idea has been expressed in ancient religious terms as the symbolic correspondence of the mundane (the earth) and the divine (the heavens), and is also represented in theories of contemporary physics which describe how each particle of matter in space contains knowledge or information about the entire system."[347]

The *Theater of Memory* (1985), an installation first exhibited at the *1985 Biennial Exhibition* at the Whitney Museum of American Art, in New York, further explored the different levels of perception as related to memory and the working of the central nervous system. A large dead tree diagonally cut through the dark room, with fifty-odd lanterns hanging from its branches. The little lanterns were programmed to flicker in unison, as if short-circuited. On the back wall, behind the tree, visible though the branches the spectator saw video images projected on the rear wall. The pictures were distorted and not recognizable, accompanied by loud static noise and sound, alternated with periods of silence. A concealed fan blew a soft wind on the tree. The installation was representative of how the artist uses memory as a function of the brain, to create a work in which he has brought images 'retrieved' from his own memory, i.e. experiences, (pre-)recorded electronic images, together with natural elements - here a dead tree which has gone through its own cycle of existence. This superposition of elements creates a perceptual experience that moves in different directions. Viola: "I remember reading about the brain and the central nervous system, trying to understand what causes the triggering of nerve firings that recreate patterns of past sensations, finally evoking memory. I came across the fact that all of the neurons in the brain are physically disconnected from each other, beginning and ending in a tiny gap of empty space. The flickering pattern evoked by the tiny sparks of thought bridging these gaps becomes the actual form and substance of our ideas. All of our thoughts have at their center this small point of nothingness."[348] Recently I was re-reading chapters of Sogyal Rinpoche's *The Tibetan Book of Living and Dying* and came across a paragraph in which - reflecting on the impermanence of existence - he thinks of a tree: "When you think of a tree, you tend to think of a distinctly defined object; and on a certain level, like a wave, it is. But when you look more closely at the tree, you will see that ultimately it has no independent existence. When you contemplate it, you will find that it dissolves into an extremely subtle net of relationships that stretches across the universe."[349] Maybe these are only my own associations. However, thoughts about the absence of an independent existence of all things on earth, called "emptiness" in Buddhist terms, seems to be a central element in Viola's oeuvre.

The Nature of Existence

Although considered a transitional work, *The Reflecting Pool* already contains the beginnings of the artist's continuing concerns with nature's cycles and his search for the roots of existence as a part of nature. A quest which he synthesized in an ambitious project completed in 1986: *I Do Not Know What it Is I Am Like.* The ninety-eight minute tape is divided into five parts, each representing a different consciousness of the the world, by taking a deep look into the psyche of the animal world, which is then related to a level of animal consciousness within human nature.[350] The title is taken from a verse in the Rig Vedas, the knowledge (veda) existing in verse (rig).[351] Central in the Rig Vedas is the uncertainty of knowledge. Questions asked here do not have answers. There is no answer to birth or death. However, this should not prevent us from asking questions, and this is in essence what Viola does all the time, in order to learn and discover. For each time we think we have found the answer to one question, we have already created the next one, as if the human being is bound by a continuous progression in the quest for knowledge, and still does not know...

Il Corpo Scuro (The Dark Body) explores the interior of a stalactite cave, before the camera moves to the vast lonely pastures of the North American bison. Both represent something of the primordial in the earthly cycles, which is brought to the fore by the long close-up shots of a dripping stalactite, of a bison's eye, of flies swarming the carcass of a dead bison. *The Language of Birds* connects in a sequel of animal portraits the eye (camera) of the artist with that of the animals, an address as well as an encounter. It links the artist's vision of the mind's eye, in both man and animal. *The Night of Sense* brings the viewer into the realm of human consciousness: we see a man (Viola himself) reading, studying, taking notes, eating a solitary fish dinner, interrupted by 'visions,' - moments of inspiration ? - moments of reconnaissance - memories? *Stunned by the Drum* tests the limitations of the eye - once again - while an intense sound of drums makes the 'eye' go wild in a euphoric trancelike state. This part, as it were, forms the introduction to the fifth part, *The Living Flame,* which takes the viewer to a documentation of Hindu fire worship, with trancelike painless dances over burning coals, and ears, noses, skins pierced with needles, accompanied by the intense sound of the drum. Yet another level of consciousness is displayed. The closing image returns to the lake where the tape began; the viewer flies in an airplane over the water with a dead fish in front of his eyes, ending with its decomposition on the shore.[352]

Over the years Viola has developed a vocabulary of images which keep recurring in different combinations with different meanings. The fish is a symbol of Christ in the Judeae-Christian religion, for the Japanese is is their main source of protein, apart from mythic connotations. In many other cultures it is more than a daily kind of nutrition. Fire and water are often used in initiation and purification rites and rituals. Light, natural or man-made, is a recurring structural as well as symbolic element. Light literally makes us see, but metaphorically means, among other things, vision, or understanding.

These are elements of the earth which have symbolic meaning in almost all cultures. Although the meaning of many Christian symbols has been lost and the same fate threatens rituals and belief systems of many of the so-called 'primitive cultures,' of which the continuity of traditional values is based on oral transmission, for Viola there are also elements which have "a natural level of meaning," implicit in themselves, and not dependent on any culturally understood interpretation. They have "a power which you sense, even if you don't know the precise meaning," he says. He calls it "the power within." A powerful example for him is the mountain, representing time itself. "The fact that we are bound together in nature, that there are certain processes happening in nature, ourselves, our bodies, our minds are part of that nature and that is the reason there are analogies happening continuously."[353]

Mind - Body Continuum

In the Middle Ages, Christian symbolism was embedded in the culture and its meaning agreed upon. Being often connected to annually returning rituals, it was also a part of daily life. So was death. In our society death is treaded as something alien, although we know that there is no escape. Mystics in Christianity, like Meister Ekkehart or St. John of the Cross, often sought to free the Self, or the mind, from the body, in which they felt it imprisoned. In eastern philosophies, though, the body and mind are not necessarily perceived as separate, both being bound in samsara. The cycle can only be broken when one has reached nirvana.

Bill Viola's work of the nineties maintains this layering and complexity of issues concerning the human experience. Yet birth and death - the two ultimate experiences in every person's life - have obtained a more dominant role. Formally it consists predominantly of video installations that include the spatial surroundings. There is a return, it would seem, to the single image, using again the monitoring quality of video. The aspect of time literally and metaphorically passing is at the center. The spectator becomes - sometimes painfully - aware of this as he or she passes through a semi-dark room or corridor, when the eyes need time to adjust in order to see.

The *Nantes Triptych* (1992) follows the medieval and renaissance system of the three-panel altarpiece. The left 'wing' shows the image of a woman giving birth, the right 'wing' of a woman dying. The central part shows a figure moving under water, symbolic for the transitional period immediately after death - already unreachable for us, yet still present? It is a continuous loop without editing. The viewer almost becomes a voyeur present at different stages in the journey of human life, which is also his own: it is the human condition portrayed. The references to the Christian triptych are intentional, yet they do not make the installation an altarpiece. Although in Christian iconography such altarpieces generally portray scenes from the birth, death, and resurrection of Christ, the *Nantes Triptych* does not carry this highly symbolic weight. It seems a reference to a medieval system of perception, in which religion, art and science were still more or less considered to exist on one plane; to a concept of nature which did not make a divison between life and death as we do. At least, this is how these ages now appear to us; 'holistic.' And this vision includes the mind in the physical reality itself.

Pneuma (1994) is a room installation in which the unity of life is fully spun out. Pneuma means breath, but also life and soul (Artistotle). Viola: "There is such a deep connection between seeing and knowing that it is odd to think how the two could have been considered separate. There has been much discussion about the result of the famous dualism of Descartes and it is obvious that today we do view body and mind as separate from each other, and we consider the intellect, senses, emotions and so on as separate parts of the mind. If we take a broader view of world culture, past and present, surveying the most advanced and sophisticated to the more fundamental and tribal it becomes apparent that the present division that we in the industrialized West have developed is a kind of distortion."[354] So in Viola's vision the five senses are not individual things but integrated with the mind; they form a total system and create this field, an experiential field which is the basis of conscious awareness.

In the five rooms of "Buried Secrets," made for the U.S. Pavilion at the Venice Biennial in 1994, the observer is acutely made aware of this. *Hall of Whispers, Interval, Presence, The Veiling, and The Greeting* challenge all levels of human perception. In *Hall of Whispers* images of ten people's heads with mouths tightly gagged are projected on the walls of a narrow and dark corridor. They try to speak but their voices sound muffled because the "gags prevent their words from being understood." The second room, *Interval*, consists of two wall size projections. On one side a man slowly washes himself scrutinously - "peaceful, calm, still" - on the other are rapidly cut images showing fire and water with a figure trying to get through and scenes of bodily fluids and orifices - "chaotic, violent, intense." Not only are we bombarded with aggression and violence day by day, from which we have to withdraw purposively, but both aspects are within us, constantly fighting. Sound, as sculptural form, is dominant in *Presence.* In this room the observer can only sense bodily presence through the whispering voices of both young and old people, "audible when (the) listener steps into (the) sound beam."[355] In *The Veiling* two projectors on opposite sides of the space project images through a series of veils hanging loosely in the room. The images are of a man and a woman nearing each other in nocturnal landscapes, crossing and moving away from each other again. The images change in clarity as they penetrate more deeply into the layer of scrims, literally become veiled. The bodies diffuse, almost disappear. Marilyn Zeitlin suggests "awareness of one's own body, of surroundings, of the narrative-line of someone else's life crossing one's own ... the multiple layers of cloth and image call up this capacity of the mind to work on all these levels at once."[356] This probably applies even more to *The Greeting*, which records a meeting and exchange of greetings between three women. The motive is based on Jacopo Pontormo's *Visitation* in the church of Carmignano near Florence. Because the episode was recorded on high-speed 35mm film, the original length was stretched twelve times. Every detail of the interchange, from the tiniest movements and gestures to the facial expressions and events in the background, therefore has equal emphasis and becomes meaningful. Yet, what is said or whispered between these women? Speech and body language are our normal human ways to communicate. They take time, literally. Communication - or rather the disability to communicate because of different obstacles, willingly or non-willingly - appears a central connecting element in these installations. Even though the viewer/listener moves through a spatial setting, the imprint of time becomes stronger than that of space. Yet many questions in this work remain unanswered - 'Buried Secrets'?

The artist acknowledges his prevailing interest in systems which deal with interrelationships between different phenomena on earth, whether on a physical or cultural, on a religious, or philosophical level. And the studies in science, religion and philosophy are no less than an attempt to bring the artist's whole field of vision together in a "total system of perception and cognition at work." We have seen that the artist's interdisciplinary research is reflected in the complex layering of a densely composed audio-visual structure, comprising among other research: - 1. research into questions of audio and visual perception, as related to the functioning of the brain, neural processing, and problems of cognition, consciousness, memory. - 2. into concepts of time and space, as related to those in the new physics as well as the electronic and digital computer technologies; - 3. into the life cycles of nature, like birth, growth, death, regeneration, as related to an underlying spiritual order in which the human body and mind are a whole.

However, Bill Viola keeps developing his work as a journey involving his own personal rhythm of life, in a close encounter with the technological progress of the medium. He chose the context of art, and one may ask how far the boundaries of art have been expanded by his vision, or in how far he contributed to a change in the definition of art.

Viola's videotapes are made for the public realm of television, his video installations are generally shown in the museum / art context. Video technology represents for Viola the only means relevant today to express these all-encompassing concerns, because according to him its principles of communication dominate the current information networks world-wide.

Chapter 12

Paul Ryan: From Guerilla Television Warfare to Ecochannel

Video for Social Change

For Paul Ryan the history of video has always been the history of a struggle between its use as a tool for social change and as a medium of art.[357] "Video itself mutated from a countercultural gesture to an art genre. When video was principally a countercultural gesture, it held the promise of social change unmediated by the art world. Now, whatever promise of social change video holds is mediated by the art world. This is a significant difference."[358] It is a distinction of degree, but it is significant with respect to the priorities being set. For videomakers who worked for social change but considered their work artistic, point one on the agenda was social change, but for videomakers who proclaimed that they were effecting social change as artists, point one on the agenda was to make art.[359] At any rate, Nam June Paik, Bill Viola and the Vasulkas have always thought that they were transgressing the boundaries of art from the point of view of the artist. 'Artist' Paul Ryan chose for the other 'non-art side,' as it were, making video works that were not coined as art in the first place. Although Ryan did participate in Howard Wise's *TV as a Creative Medium* (1969) with a piece titled *Everyman's Moebius Strip*, which was also his first show in an art gallery, he admitted later on to have been reluctant to commit himself, because his video came before art. Howard Wise had called him at the suggestion of Nam June Paik, whom he had met through one of the artist's *Electronic Art* exhibitions at the Bonino Gallery.[360] *Everyman's Moebius Strip* was, Ryan wrote, "Essentially a booth in which people could experience their video image in private, knowing that the tape would be erased."[361] Central to the work was the idea of video as a system of communication, albeit confined to the relationship between the viewer and the system, between the viewer and his 'mirror' image. In addition, one can say that it also dealt with the implications of the power which the videotape gives you to see yourself as others see you.

Although I have already mentioned the exhibition in the chapter about Nam June Paik, it is worthwhile to note that only two installations, Frank Gillette and Ira Schneider's *Wipe Cycle* and Paul Ryan's *Everyman's Moebius Strip*, as well as Nam June Paik and Charlotte Moorman's performance, dealt with the external communications aspects of video as a two-way system, instead of only its internal qualities as an image processing system. *Wipe Cycle* consisted of nine monitors, a TV camera and videotape. Using feedback, the spectators could see themselves immediately when entering the space, mixed with pre-recorded images and a normal television program. The images moved in four cycles from one monitor to the next, with an eight to sixteen second delay. Ira Schneider explained to Jud Yalkut: "The most important function of *Wipe Cycle* was to integrate the audience into the information. It was a live feedback system which enabled the viewer standing within its environment to see himself not only in time and space, but also eight seconds ago and sixteen seconds ago. In addition he saw standard broadcast images alternating with his own delayed/live image."[362] Frank Gillette started working with video in 1968, coming from painting. *Wipe Cycle* was his first major piece made in collaboration with Ira Schneider. Marshall McLuhan is said to have given him the opportunity to utilize a portable video recorder. Gillette became known for his videotapes and installations 'depicting' biological processes in nature, using real time and time delay devices.[363] Ira Schneider had studied art-history, but turned to filmmaking. In 1969, he began working in video, and he has made a number of video installations.

To recall, most artists of the early hour utilized such technical features as feedback and closed-circuit as a kind of 'self-communication.' A small group, some of its members coming from music, like Ron Hays and Steve Beck, was interested in computers, and experimented with video/television's image processing features. And those artists who connected video with social change and its educational possibilities for cable, emphasized the communication aspects and discussed video in terms of an information system. The latter generally recorded in a rather straightforward manner.

Paul Ryan, Frank Gillette and Ira Schneider had all become involved with Raindance Corporation, which was founded in 1969 by Michael Shamberg and others as a countercultural think tank. Guerrilla TV groups like Raindance started out with ideals about changing TV's one-way communication system into a two-way system which would allow the viewer to participate in the programming, after Marshall McLuhan's ideals, seeing television as a future global communication system. Ryan: "We made tapes as a collective and showed them at a loft on East Twenty-Second Street in Manhattan. Some of these black-and-white 'raw' tapes are still extant in the Raindance Archive."[364] Among the tapes from that period were a documentation of 1970's Earth Day in New York; *Rays*, a tape made on a California beach; and material which dealt with surveillance video then used in supermarkets. Raindance was one of the many video groups at the time which were predominantly concerned with presenting pure information. The Alternative Media Center, Global Village, Video-freex, and People's Video Theatre were other groups that tried to invent new ways of using video to mediate social conflict.[365] Their interest was above all social and political. Raindance Corporation purposively set out to develop alternative approaches to the television structure. With this in mind they videotaped political rallies and unions, protests, social injustice, street life, and so forth, and also published *Radical Software* (1970-1974). At the time, Paul Ryan was convinced that portable video presented a true weapon for cultural transformation. In *Radical Software* (Spring 1971), he announced that the concerns of portable video were comparable to those of guerrilla warfare, in which the guerrilla fighter generally moves invisibly, and attacks by surprise and in unexpected places, because he knows the terrain. Ryan looks back at this text as a bit of an 'angry text,' written in a rhetoric that was mostly determined by his anger about the Vietnam War.

Practically, the 'guerrilla warfare' groups saw their chances with cable television (CATV), which was rapidly becoming a major business in the early seventies. The point of view was that cable television's business lay in the cultivation of local culture, and that the role of a cable system was to increase the community's awareness of their existing local cultural system, whereby the possibilities of its public access would give more power over the local situation, i.e. cable would allow people to have input, extend control over local politics, local culture, etc. Ryan commented: "Just as VTR extends man as a cybernator so cable can enlarge the capacity of the local culture to communicate about and control its development. This control can include some decisions about importing information. ... Low priced portable video units make it possible for the cable company to take their whole district as their studio. Feeding back into the culture rather than feeding off it will insure lasting relations between cable and culture."[366] For Paul Ryan it went without saying that the pervasive influence of broadcast television had to be counteracted and that video had come right in time to function as a tool of protest. The 'video belief system' of this group can be deduced quite clearly from the editorial statement written by Michael Shamberg for the *Radical Software* issue on "Guerilla Television," and it is worth quoting: "In issue one, volume one of *Radical Software* (Summer 1970) we introduced the hypothesis that people must assert control over the information tools and processes that shape their lives in order to free themselves from the mass manipulation perpetrated by commercial media in this country and state controlled television abroad. By accessing lowcost 1/2" portable videotape equipment to produce or create or partake in the information gathering process, we suggest that people would contribute greatly to restructuring their own information environments."[367]

Paul Ryan switched to electronic technology in 1967, giving up a career as a writer, in order to avoid more Vietnams, following the idea put forth by McLuhan that the new electronic communication media would bring a more harmonious society. In 1966, Ryan had been drafted to go to Vietnam. As an activist strongly opposed to the Vietnam War he asked for alternative service, and his request to work at Fordham's Communications Center where Marshall McLuhan was appointed visiting professor (1967-1968) was granted.[368] There were a number of aspects in McLuhan's thinking that fascinated him. Being a writer, he was intrigued by McLuhan's disregard of linear sequence and the fact that he quoted at will any reference at any time to make a point. (Note that the writings of Robert Smithson also showed this tendency.) However, perceiving similar qualities in electronic technology, he started to experiment with portable video equipment and tried to use this new medium to fashion a non-linear way of thinking. As an (almost) political activist, the economic and political implications of portable video were self-evident to him. The idea was that, because video was a self-contained system, it could deal with culture's processes by itself, whether family, classroom, therapy, or cultural issues at large. In opposition to television, it allowed one to raise issues that seemed important to the individual person or the group, etc. As a consequence of the controversial information that was broadcast via the networks, and considering the conflicts that arose about American foreign policies during the Vietnam War, people began to doubt that they were being told the truth about the death of relatives, for example. This questioning of the information led implicitly to the following questions: what else is whithheld, by whom and why? How is this system of communication maintained by those in power? And how does power function ?

If the new portable video system was seen as a practical instrument to get access to controlling the flow of information by means of 'guerrilla warfare' systems, some kind of theoretical foundation was wanting, if one wanted to be successful in countering the established structure of the television networks at all. Consequently, Ryan set out to define the utilization of the new medium video in relation to the cybernetic and media theories of Norbert Wiener and Marshall McLuhan, what he thought relevant in their ideas: "Cybernetics understands that power is distributed throughout the system. Relevant pathways shift and change with the conditions. ... Cybernetic guerilla warfare...because the tool of portable video is a cybernetic extension of man and because cybernetics is the only language of intelligence and power that is ecologically viable."[369] In the meantime, he had met Gregory Bateson at a conference in 1970, and his ideas that the cybernetic circuit as the basic unit of mind could be identified with the basic unit of evolutionary survival had impressed him deeply.

An Evolutionary Tool for Education

Since then Paul Ryan's video works have taken a distinct direction. Starting with video as a medium for personal change, as a medium that allows one to experience personal power, he subsequently moved to a level where this 'power' is used for social change. This is the road that he has consistently pursued to create a body of theoretical writings and video works that involve the creation of a new "information-transmission system based on shared perceptions of environmental realities rather than on language."[370] Ryan's perception of video as an evolutionary tool for social change had been enlarged by the notion that this change should be accompanied by a growing ecological awareness. And these concerns about the earth and its ecology have remained central issues in his work; in fact they became the content of his work. The concern for the earth and its ecological systems was, interestingly, also a dominant feature of the contributions of *Radical Software* between 1972 and 1974.

It became clear to him that if he wanted to make some of guerrilla television's goals come true, it was necessary to seriously investigate the roles which the 'new medium' video could play in education, as Nam June Paik had already indicated. Ryan, coming from a group that sought ways to produce alternative television, first thought of the possible educational functions of cable television. It became a central element in his future concepts. Only, the cable industry had to be convinced of the necessity of this function. Although he has expanded his visions to include other environments as well, a major part of his work now consists of the development of a teachable method for cable-TV or television that will put forth "video interpretations of the natural world" in view of a communication that takes into account the complexity of the earth's ecological systems.[371] The development of his thinking, he says, was basically formed by the semiotic theories of Charles S. Peirce and by Gregory Bateson's application of cybernetics in the social sciences. Peirce is considered the father of American semiotics. He became known for the development of a logic of signs, which proposes a scientifically formulated classification system of the sciences. It is in essence a logic of relationships, which researches the system by which objects are created and connected by relationships of any kind.[372] Peirce's starting point is that all living human beings share the same world, and that they (we) carry the responsibility for this world. For human beings are the only living beings who can consciously decide what to do and where to go, and decide how this world is going to look. This common experience allows us to share information, or knowledge, and thus to develop a common perspective for understanding. Gregory Bateson's research has focused on the nature of man and his relationship to the environment. His interdisciplinary studies, whether in his own field anthropology, or linguistics, biology, perceptual and behavioral psychology, eventually all aimed at creating a classification of contexts of behavior by applying cybernetic principles. Bateson proposed, among other, that all biological and evolving systems, that is individual organisms, animal and human societies, eco-systems..., "consist of complex cybernetic networks, and all such systems share formal characteristics."[373] His wide-ranging interests also led him to research the nature of learning processes, such as biological adaptation, or the acquiring of knowledge, and its connection with evolutionary processes. Why is one idea at a certain time given preference over another, for example. Is there indeed some natural selection process at work here? Bateson maintained, in contradiction to Darwin's theory of natural selection, that the unit of survival was not the family line or the (sub)species, but "the unit of survival is *organism* plus *environment. ...* a very strange and surprising identity emerges: *the unit of evolutionary survival turns out to be identical with the unit of mind.*"[374] Since in his vision the mind is immanent in the large biological system or eco-system, and biological structures (living organisms) and grammar (as a product of mind) can both be classified as products of a communicational and organizational process, Bateson deduced "that this massive aggregation of threats to man and his ecological systems arises out of errors in our habits of thought at deep and partly unconscious levels."[375] Basically from these concepts Paul Ryan evolved the idea that it must be possible to make conscious the uncon-sciously shared levels of perception about the ecological environment.

Inspired by Warren McCulloch's concept of *calculus of intention*, Ryan began research on communication and behavioral patterns among groups that consisted of three people (1971-1975). Warren McCulloch was a mathematician/cybernetician at the Massachusetts Institute of Technology who had specialized in research on brain circuits. He had come to the conclusion that communication in a dialogue largely broke down because human beings lacked a logic of handling relationships in group of three or larger, called "triadic relationships."[376, 377] Paul Ryan began a series of videotapes in which he experimented with people's interactions in sets of three. From there he developed the concept of a "triadic logic" as a relational practice among groups divided in threes. Via this relational practice, Ryan thought that it might be possible to build a communications system based on a shared perception of the environmental realities which we live in. The only thing needed, according to him, was to make people aware of this shared perception. This triadic logic was to have a direct educational purpose. Concerning the videotapes, he said: "I trained myself to do half hour continuous tapes of the environment in a Zen state of mind. I did a set of 36 such tapes dealing with earth, air, water, plants, animals and man, considered in his technologies" (1973-1976). He had learnt the practice of T'ai Chi and meditation. The result of this period were *The Triadic Tapes*, 12 hours of black-and-white videotape, and *Earthscore Sketch*, 18 hours of black-and-white videotape, divided into 36 continuous half-hour segments, both of which were shown on Manhattan Cable, and presented at the Kitchen's Performance Space (1976). In 1975, he did a year-long study of a waterfall in High Falls, New York, watching waterflow patterns, which he edited down from thirteen hours to an hour-long videotape titled *Water Chreods* ("chre = necessary; "ode" = path in Greek). Chreods are nature's basic pathways or patterns. By systematically observing the patterns underlying a particular eco-system, Ryan set up a vocabulary of chreods which "can give us an articulate set of notes with which to 'score' natural phenomena. ... The syntax of interrelationships between these chreods would, in effect, constitute the 'score' for that particular system."[378] The artist does not emphasize the themes recurring in his work. He prefers that his work not be dealt with thematically, as he feels it might harm the idea of a (dis)continuous evolutionary development. However, it is worth mentioning that the artist has repeatedly returned to the theme of water, "as the richest single source for developing a vocabulary of 'chreods' in nature. ... Water takes so many different shapes - billows, droplets, backcurls, waves, fantails, and cascades - and each of these shapes exhibits a different pathway in which water can flow, a different chreod..."[379] In 1983, he did a study of the Great Falls in Paterson, New York, which resulted in the videotape *Where the Water Splits the Bank* (1985). It was followed by a study of the coast of Cape Ann, Boston in 1984, with a tape of the same name, *Coast of Cape Ann.* In 1986 Ryan crossed the Atlantic Ocean in a North Sea trawler, taping over 30 hours of ocean waters. Yet another videotape, entitled *Mountain Waters,* interprets different water ecologies in the Shawangunk Mountains north of New York City. The last one in this series, *Water Fire Water* (1993), relates sets of water 'chreods' and fire 'chreods.'

Paul Ryan denotes the time that he lived in the Shawangunk Mountains (1971-1975) as his early earthscore period, as the time in which he found a formal basis for using video as a means to evolve and develop new learning patterns for human behavior and perception. "The name Earthscore indicates my assumption that ecological systems, like musical systems, are made up of 'notes' and that humans can interpret these notes with video." The Earthscore model is, among other things, based on the principle that people can and do process information according to the uniqueness of their perceptual systems. He connected the cybernetic concept of self-correction (a classic example is the thermostat, or the temporature of the human body) with Charles Peirce's concepts of firstness, secondness, and thirdness. Firstness is defined by Peirce as there being no regard for the other, secondness as correspondence to the/an other, for example, there is a reaction, and thirdness as the correlation between firstness and secondness: there is mediation, or discourse. After experimenting with his own triadic model that he had set up using the video as tool of communication between three persons, so that communication was thus defined as a set of interacting patterns of behavior, Ryan came to the following conclusion: "I think that there are two interrelated possibilities: the invention and maintenance of a repertoire of behavior patterns that would work self-correctively for three people and the construction of an information system based on shared perception of environmental realities. ... In terms of the self-corrective triadic behavior, what it seems possible to develop is a 'practice' in the Oriental sense of the word. In the Occidental sense, one practices in order to acquire a skill, e.g. to be able to shoot an arrow straight. That skill is then a new tool, which you, unchanged, now have. In the Oriental understanding, you practice in order to change yourself. By incorporating a discipline relevant to archery into yourself, you become, out of practice, a different person... three people could incorporate video and video feedback into their interaction and change the 'normal' pattern of three party differentiation into an ongoing self-corrective process. The people themselves would also change."[380] This quotation shows how Ryan goes about connecting an idea from Peirce's semiotic system and from Bateson's cybernetic notion of a self-corrective circuit applied to human behavior with his own model of triadic logic, with the intention of creating a shared perception of environmental realities and a view of the ecological environment as a whole, including man and nature, as the condition necessary to accomplish social and cultural change. It is also the key to the further development of his Earthscore Notational System. Ryan: "The Earthscore Notational System grew out of my efforts to use video to interpret nature ... to produce an orchestration of perception. ... In effect the Earthscore Notation provides the architectural plans for developing an information transmission system based on shared perception of the natural world and not speech or writing."[381]

The Earthscore System as Video-Ecology

In 1977, Paul Ryan went to San Francisco to meet with a group called the Frisco Bay Mussel Group. This group stood for bioregionalism, which can be defined as an involvement that cultivates strategies for sustaining living in the context of local ecologies. The group's involvement was above all political and by that time had exchanged its early utopian community ideals for a functional way of operating to achieve its goals.[382] At the time the bioregional approach was new and important, because it viewed the eco-system of the bay as a whole. Ryan was particularly interested in 1. the practical view of developing strategies to sustain what was called "a culture of place," and 2. the theoretical foundations of bioregionalism, offering a possible connection to cybernetic theory.[383] Meanwhile he had also learnt about James Lovelock's Gaia Hypothesis. Lovelock perceives the earth as a living organism, not in the metaphorical sense, but cybernetically. According to Lovelock, the phenomena that determine the 'organic' balance on earth, such as atmospheric conditions, like temperature and oxygen/nitrogen ratio, are based on self-corrective feedback mechanisms on a planetary scale.[384] The Gaia Hypothesis suggests that we have - or can acquire in a reasonable frame of time - an operative knowledge of these planetary mechanisms, and that we can learn to identify them, and thus eliminate errors that we have made related to processes in nature, or planet earth. It is not a free ticket for those who might think that the earth will therefore adapt to whatever circumstances. Lovelock has been criticized for the theoretical possibility of the implication. On the contrary, his theories really imply a new ethical understanding, as opposed to the position, still often held, that science is value-free. The Gaia Hypothesis - science has had to accept at least a few of Lovelock's theses - reminds the scientist-researcher to keep asking how far one can go before the self-corrective mechanism can no longer return to its original position.

In 1979 Paul Ryan moved to North Jersey again, and decided to start *Talking Wood*, a magazine advancing these principles.[385] In order not to remain stuck in theoretical rhetoric about the destruction of the environment, such as 'if we destroy our environment we will destroy ourselves,' the organization initiated a Watershed Watch, which relied on Ryan's concept of "a shared perception of the watershed by the people living here, ... suggesting a (new) way of coordinating our natural watchfulness of the place we live in, connecting multiple aspects of the site methodically and relationally, as circuit."[386] *Talking Wood* was structured as a public activity, using Peirce's categories and the notion of a relational circuit as design concept, to make people aware of the various ecological elements of the Passaic River Watershed. The organization's strategy was to work directly with the local population, through workshops and lectures, identifying problems concerning the eco-systems that deserved attention. Objects of study were among others the flood problems in the Passaic River Basin, and developing a program for making the city of Paterson a so-called 'green city.' The Passaic Watershed Watch set up an interdisciplinary design group to do an engineering study to meet flood control and associated water-related land resource needs of the people of Passaic River Basin. The use of video and television as educational tools were central elements in the proposals and workshops, being considered, of course, the most effective contemporary means of communication. The positive results of some of the projects undertaken were an affirmation that at least small groups of people were now taking responsibility for the public habitat. These activities can be seen as part of a more or less continuing development toward yet another 'shift' in activities, as Ryan began to work full scale on the design of a 24 hour television channel dedicated to monitoring the ecology: the *Ecochannel.* Again, it fell together with a move, this time to Hoboken, New Jersey. The thoughts for this project were developed between 1981 and 1985.

Design for an Ecochannel

The proposal for the *Ecochannel* itself was concisely described: "The Ecochannel is a design for a television dedicated to monitoring the ecologies of a region and developing consensus among inhabitants about sustainable policies for that region," using the model of the Earthscore Notational System.[387] It is based on a systematic mapping or "inscaping," as Ryan calls it, of the natural patterns of a bioregion, in this case New York, using videotape. It involves the recognition of wants and needs, and foremost of the shared perception of the environment, as explained. The application of the computer has made it possible to encode the givens by transcribing the data into mathematical models. Ryan envisions that the method can be expanded beyond the local educational forum in schools and community centers to include telecommunications media, such as satellite, cable television or computer networks. The *Ecochannel* design has thus far been presented at the Museum of Modern Art in New York, the First International EcoCity Conference and at the United Nations.

The focal point in the proposals, or 'designs,' which Paul Ryan has been developing during the past five or six years, is the development of an ecological consensus, which is in fact a continuation of his earlier premise of the creation of "shared perceptions of environmental realities." The *Design for Ecological Consensus along the Hudson* has as its overall goal the achievement of an ongoing consensus about the ecology of the Hudson River among the parties interested, the regulatory agencies responsible, and the general public. The starting point is again to create a design that will lead to a perceptual congruence of the whole environment as an ecological system. A communication system will be created to discuss what it means to live in accord with the ecology of the Hudson on a long-term basis.

The design - also called Environmental Mediation - is set up to provide a context that would make it easier to talk about and handle disputes in conflict situations between the body politic, environmental organizations, the industry and other interest groups, as well as the public at large. Environmental Mediation is set up as follows: 1. it will use a neutral third party - thus based on the logic of triadic relationships -, to formally create a formal neutral set of relationships, 2. it is based on a jury of observers which may preclude advocacy in favor of one or the other party, 3. it is set up to create a circle or circuit of understanding which "could produce an architectural basis of an information management system..."[388] Ryan's vision is that the method could be embedded in a computer program of a database network system that would help orchestrate communication, in this case about the Hudson ecology, and make that communication available to anyone with a personal computer. What it means in terms of educational practice is that the whole ecology - or at least as many aspects of the ecosystem as possible - may be used as a common reference for different subject areas taught, including science, mathematics, literature, social studies and art. Being an interdisciplinary and experimental method, Ryan explains, it comprises four components: 1. the ecological setting itself, 2. the learning about ecology by facts, patterns and context (Peirce's firstness, secondness and thirdness), 3. the system of knowledge already available, and 4. as a cooperative learning procedure, he hopes that it might ultimately create the 'new' way of thinking that Gregory Bateson had envisioned as "the ecology of mind."

Among his recent proposals is the New York City Environmental Television Channel, called NEST (New York City Ecochannel for a Sustainable Tomorrow) with the purpose "to monitor and interpret the eco-systems that support New York City, so the citizens of New York can develop ecologically sound policies and practices."[389] In answer to the question why he considers television the appropriate medium to achieve his goals, Ryan says, that for him television functions as an "electronic window on city ecologies and thus can develop a shared perception of how these ecologies work, as well as an understanding of how not to destroy them. ... Television, because television enables us to monitor events simultaneously with others, and this is what tv does best."[390]

The artist cofounded and directed the Gaia Institute at the Cathedral of St. John the Divine in New York City, and has also joined the Earth Environmental Group, a non-profit organization dedicated to environmental concerns at large.[391] The need for environmental education is urgent and has to be far-reaching. An example of work related to these theoretical and educational methods is *Nature in New York City* (1989). In this videotape the spectator is taken to four different natural habitats. Site one is Jamaica Bay, the Gateway National Recreation Area where the horsehoe crabs lay their eggs. One first sees the laying of the eggs, signifying the forthcoming birth of new crabs, and thus the procreation of the species. The artist explains this as Peirce's firstness. This process can be interrupted by predators, i.e. animals, human beings (Peirce's secondness). Finally, we learn that it is part of a pattern, including the crabs' life cycle, as well as the human/urban environment (Peirce's thirdness). Site two shows Clay Pit Pond, at Clay Pit Pond State Reserve, Staten Island. Here, the artist selected five phenomena present at the site: deciduous trees, evergreen trees, abandoned cars, grass and reeds. The spectator's 'eye' sees the surface of the pond and aspects of its surroundings to get an impression of their qualities (firstness). This section is followed by a more detailed description, functioning as factual information of the five phenomena (secondness). Finally, a visualization of patterns and relationships between vegetation, human intervention and the pond is created (thirdness). Site three shows a group of trees in Inwood Hill Park forest (Manhattan, NYC).

Ryan videotaped the natural phenomena 'pure,' such as the melting of snow, the barks of the trees, green elements (firstness). The viewer is then confronted with indications of the presence of men such as burnt wood, litter left (secondness), and the tape ends with children playing among the trees (thirdness). Site four presents a waterfall in the Bronx River, at the New York Botanical Garden in the Bronx. Here the artist focused first on the water itself, its surface qualities (firstness), then on the turbulance, the structure of the water as a waterfall (secondness), and thirdly on the explicit water patterns and the geological context of the falls (thirdness). On a compositional level, *Nature in New York City* uses what Ryan has referred to as "the relational circuit," an organizing mode of composing the knowledge obtained according to the categories of firstness, secondness, and thirdness. The purpose of Ryan's notational system is that it helps us to identify errors in our treatment of an ecosystem, by learning to understand the underlying patterns. For example, the crabs lay their eggs in the wet sand during the extended ebb tibes when there is a full moon in June. This way maximum protection for the eggs during the birthing process is assured. But when this "figure of regulation, a chreod, is destroyed, the natural process is destroyed."[392]

Paul Ryan has moved from 'guerrilla warfare,' whose tactics are secretive and take one by surprise, to a strategy that includes open discussion. He has taken up a position as an artist, not as a social worker, or environmentalist or ecologist, although his work covers large parts of the territories of these professions. In this, his position resembles Alan Sonfist's, who has moved into the territory of the landscape architect, and who is now often called an ecologist/artist. His approach is that of an artist, not of an educator, although he spends most of his time teaching. The "threeing" practice has been successfully applied by Ryan in art and educational programs since 1976. Recently he has used the Earthscore Notation and specifically the triadic logic to design and train others to teach a course for workers displaced by the defense industry, for ETI, a 12 million dollar employment and training company. Paul Ryan has developed Earthscore into a "method for generating curriculum for education for sustainability; design of learning experience, 'composing' my art in a non art context."[393]

Many artists in the United States hold teaching positions, but for most of them it is just a job, a way of earning a living. For Ryan, education is part of his art work, if not the central part. Again, for him education is not restricted to what we usually think; it includes a new approach to teaching, which one might define as an interdisciplinary systems approach, with the intent to create an ecology of mind. His central theme is the ecology of the earth, running as a continuous thread through his video work and thinking. One may compare Paul Ryan with Helen and Newton Harrison, as they share similar concerns, and whose work also has an educational component. The Harrisons also work on an interdisciplinary systemic level, taking into account both nature's eco-systems and human conditions. Yet the Harrisons have presented the major part of their work in an art context, Ryan rarely has. Paul Ryan will show his work in museums and galleries upon invitation, but he is not at all interested in art politics or its commercial aspects. After all, he is not making art objects that can be put on display. The much broader context of society is his domain. In essence, one could place Paul Ryan among a new category of artists who have been defined by Gene Youngblood as metadesigners, and who set out to create new contexts to replace or function as an overlay upon existing ones.

Chapter 13

The Vasulkas: Video Phenomenology

Inside Electronic Media

The artists that have interested us, have connected the aesthetics of the internal frame with the external communication aspects of these tools, like Nam June Paik, Bill Viola or Paul Ryan. Also, they have perceived their work as a discourse between video as art and video as television, and/or as exploration of the communication structure of the medium. Yet another direction evolved from the intersection of the communication technologies and the electronic media arts, one that is characterized by research into the possibilities of the inner technological-aesthetic workings of video. This has been the path of the two artists Steina and Woody Vasulka, also known as the Vasulkas. Their aim was from the start to advance an understanding of the inner and outer workings of electronic media, both aesthetically and technically, and to advance awareness of the context in which these technologies operate, and possibly have an influence on that context.

Among video artists the position of the Vasulkas is quite singular. The work of 'pioneer' video artists Steina and Woody Vasulka can be characterized by a continuing inquiry in the electronic processing aspects of the medium 'video,' from an innate desire to understand at first the inner workings of electronic phenomena, and later of digital ones. In so doing, the artists have not only made a major contribution to art, but also to the development of image processing. They are the only artists who have persistently sought to integrate the matrices of the 'system,' like the video signal itself, in the creation of images and how these could have a meaning in themselves. Since the introduction of the digital editing system and the digital computer, their inquiry has expanded naturally into the syntactic elements of these media.

Woody Vasulka was born in Brno, Czechoslovakia, 1937, as Bohuslav Vasulka. He studied hydraulic engineering, in which he obtained a bachelor's degree in 1956, but decided to change careers and enrolled in the Film Department of the Art Academy in Prague, where he graduated in 1964. Steinuun (Steina) Briem Bjarnadottir was born in Reykjavik, Iceland in 1940. She studied the violin and music theory. In 1959, she obtained a grant to continue her studies at the Music Conservatory in Prague. The two met in 1962, married in 1964 , and moved to the United States in 1965. Since then they have worked together collaboratively and independently. Contrary to the Harrisons, the Vasulkas work more independently, especially lately, acknowledging both the shared contributions as well as the differences in their approach.

When Woody and Steina arrived in New York City, inter-media art was at its height. Diving into an active New York art scene, they found themselves attracted by the experimental side, to artists working in the margins. During the first two years, Woody Vasulka returned to film - the medium he had worked in back in Czechoslovakia. In the summer of 1967, while working on multi-screen films, he met Alphons Schilling, with whom he began to experiment with electronic sound and stroboscopic light. In 1969, they met Harvey Lloyd, who had built a small cheap little studio where he had begun to make videotapes, which Steina frequently watched. These early videotapes still used video feedback as imaging code, but, as a trained musician, Steina became especially interested in the sound - image relationships. Woody, however, was immediately struck by "the image representing an energy system," to use his own words.[394] It was the beginning of "their dialogue with the machine." "I have to share the creative process with the machine. It is responsible for too many elements in this work," said Woody.[395] Note that this is a different attitude from the one that sets out to be the master of technology, to 'master' nature. Thus from the start the Vasulkas were not so much fascinated by its images, as by the capabilities of electronic technology itself.

Setting up their own studio in 1970, they bought a portapak, as well as a sound synthesizer and three monitors. Woody Vasulka had left his job and, in order to support themselves, they now had to apply for fellowships and grants. As New York State Council on the Arts funding was only available through non-profit organizations, they founded the group "Perception," together with Juan Downey, Frank Gillette, Beryl Korot, Andy Mann, Ira Schneider, and Eric Siegel.[396] In 1972, they already received funding from the National Endowment for the Arts and were invited to participate in the programs of KQED-TV's National Center for Experiments in Television (San Francisco) and WNET/Thirteen Television Laboratory (New York). In the meantime they had co-founded The Electronic Kitchen, later on The Kitchen, as a center for electronic music, video and performances. The space was set up as an informal laboratory for artists to meet and experiment with sound and inmages, and would become *the* center for these activities during the seventies and part of the eighties.[397]

Woody Vasulka summarized the early excitement about the "new medium" and all the activities that evolved from it as follows: "What is special about video art at this time (1972) is that it isn't yet trapped in rigid rules. There are not yet any clichés, and the artists haven't had time to develop the maniacal egos one finds in the other arts. All the video artists are like one big family and thinking about video's big future."[398] The Vasulkas were interested in art and counterculture, but less in politics. However, counter-culture and politics were closely allied then, so they hovered for a while between the 'established artists,' exhibiting in the art circuit, and the politically active 'community art' groups. Steina Vasulka described their position succinctly when she said: "None of them (by which she meant groups like Raindance or Global Village) were particularly interested in art, although a lot of them had art backgrounds. ... This was their anti-art statement, so that set us immediately on the fringe, because we were never really interested in politics. I saw it as an American internal affair that was very interesting for me to watch as a foreigner, nothing else."[399] For the Vasulkas initially these roads crossed. While experimenting for 'artistic' reasons, they also made documentaries for the Alternative Media Center in New York and, like Nam June Paik, compiled an archive documenting this cultural world of New York City.

Digital Space as Perceptual Environment

Like Bill Viola, Woody Vasulka emphasized the fact that the electronic image does not have a single focal point, being a wave (or particle) pattern of electrical energy, electrical impulses consisting of voltages and frequencies, always in motion, and dynamic. Because video does not have a fixed perspective, it need not conform to traditional forms of representation of figuration, according to their point of view. Although the Vasulkas are interested in how images are seen or perceived as reality, their premise was not the eye of the camera recording reality. The Vasulkas soon took to an invisible reality as a starting point: the internal time/energy parameter of video technology as an "inner model of imaging." Their research into the medium itself started as 'the tool' from which an image is derived. To get a better idea of their approach it may be helpful to set it against Bill Viola's. Viola starts with a "question" or an image in mind and then develops the necessary technological tool for it. In addition, Viola models a visual syntax through the eye of the camera, in order to create a perception of reality as "the mind's eye." Woody and Steina Vasulka describe their method as a "dialogue with the tool and the image, so we would not preconceive an image separately, make a conscious model of it, and then try to match it. We would rather make a tool and dialogue with it. ... But it is more complex, because we sometimes design the tools, and so do conceptual work as well."[400]

Woody's initial fascination centered on the differences between film and video; video has no single frame, for example, film does. Also the electronic spectrum is not confined to a frame but exists in a principally undefined space. Video space is a floating space. Steina described her first encounter as falling in love: "As soon as I had a video camera in my hand - as soon as I had that *majestic flow* under control, I knew I had my medium."[401] Their research has from the start included both the nature of the video signal and the audio signal, these being based on the same electronic principles. Sound can be used to generate video images, and vice versa, as the early experiments with audiosynthesizers proved. Many of their early videotapes and 'environments' are experiments in this relationship.[402]

The investigation into the matrix of video led to a series of works in which images and sounds came purely from the machine. A next step consisted of exploring the nature of the spatial aspects of the video image, and in particular the horizontal drift, in which images literally drift horizontally from one monitor to the next. A three-segment videotape from 1970, *Evolution*, 'organizes' the various stages of human evolution using this feature. At the time they could not yet technically control the speed of the drift. This problem was solved in 1972, when they had a Horizontal Drift Variable Clock built for them by tool builder/engineer George Brown. In *Spaces I* and *Spaces II* (1972), the horizontal drift is brought together with video-activated sound imagery. *Spaces II* textures three layers of shapes, whose image planes move simultaneously and horizontally over the screen. The next step was to create images in which foreground and background relationships were manipulated, exploring spatial effects and the three-dimensional possibilities. So, Brown constructed a so called multi-keyer for them, which made it possible to layer up to six images on top of each other and key them to manipulate the images at will (the normal standard at the time was two). The Vasulkas described the numerous videotapes that resulted from the increasingly complex combinations as *Electronic Image and Sound Compositions*. The images of these videotapes are composed of different wave modulations which changed from resembling abstract forms to and organic ones; sometimes they even resembled landscapes or created other associations with nature. "They resemble something you remember from dreams or pieces of organic nature, but they never were real objects, they have all been made artificially from various frequencies from sounds, from audible pitches and their beats."[403]

Woody Vasulka once explained why and how they decided to take this direction of not accepting the available hardware and the existing possibilities it offered. Being educated in film, he had come to understand the economic structure of its production system - the studios, laboratories, equipment and so forth - as a system in which the filmmaker was subordinate to the existing structure. When he started working with electronic equipment, it was for the first time that he managed to achieve some independence, or, as he called it "to personalize the process of image-making." This personalization of the tools has become their consistent striving. In a sense the tool has come to equal the medium. Of course, economics played a crucial role as well. One has to remember that before portable video equipment and the personal computer became consumer goods, these technologies were mainly utilized by reasearch facilities and laboratories. However, Woody Vasulka came across one feature of the American industries that fascinated him tremendously, as he recalls: "In working with electronic systems, I've been able to observe how they became available, how they filtered down from this industrial or commercial world to the point where they were within my reach. I also discovered that in the United States there's an alternative industrial subculture which is based on individuals, in much the same way that art is based on individuals. ... These people, the electronic tool designers, have maintained their independence within the system. And they have become artists, and have used the electronic tools which they had created. We've always maintained this very close symbiotic relationship with creative people outside industry, but who have the same purposeless urge to develop images or tools, which we all then call, maybe, art."[404] Thus it happened that Vasulkas began to collaborate with a community of engineers and artist-engineers like Eric Siegel, George Brown, Steve Rutt and Bill Etra, Don MacArthur, and still do, in order to develop the tools they needed for their work.

When the Vasulkas purchased the Rutt/Etra Scan Processor in 1974, a video-synthesizer, Woody felt that this piece of equipment gave him a wholly different understanding of the electronic image, since it enabled him to perceive the image as a product of time and energy. Rutt/Etra processes a signal so that the electron beam, that is light energy, scans an image in such a way that it becomes a series of waving lines. A face, for example, dissolves into a kind of topographical map where the lines indicate high and low areas. The bright areas are raised and the dark areas lowered; a three-dimensional image appears, because the time segments evolved step-by-step from two-dimensional to three-dimensional patterns, using the wave form as a programmable building element. Or, as Scott Nygren explained: "The Rutt/Etra reorganizes imagery by electromagnetic deflection of the electron beam; deflection coils form a yoke surrounding a monitor built into the synthesizer. The resulting images can only be recorded by means of an external camera, since the wave form display, or raster image, alone is altered, not the wave form code directly."[405] For Woody this feature became a central element in creating a syntax for electronic images, at least for a while. For Steina it functioned as a deconstructive device. The Rutt/Etra Scan Processor not only allowed the video raster to be reshaped through magnetic deflection, it could also visually display the most basic elements of the video signal itself in a precise manner. It also helped in clarifying how the conversion of light into energy happens sequentially when the electronic image is created; the time frame attained particular significance. It confirmed Woody's notion that video was in fact nothing more than electromagnetic energy structured in time. Videotapes made between 1974 and 1977, such as *Reminiscence* (1974), *Telc* (1974), *The Matter* (1975), and *C-Trend* (1975), try to depict this process.

Woody: A Code for Electronic Images

Gradually Woody Vasulka's interest in realizing images without the organizing principle of the camera as the mode of electronic image forming increased. The departure from the "camera-obscura principle" and its visual/perceptual references to an outside world seemed the next logical step, as his analysis comprised smaller and smaller time sequences in order to understand how wave formations worked in as many components he could possibly program. This departure coincided with the development of the *Vasulka Imaging System* in 1975, a digital computer-controlled personal facility. This system allowed Woody to make the shift from analogue wave form patterns to digital computer-generated structures. As we know, the binary code, 0/1, is at once a mathematical notation and a formula describing the signal code. It organizes, as it were, the language of the computer. For Woody Vasulka, the non-camera-derived images, solely created through the arithmetical and logical unit (ALU) of the computer itself, entailed the possibility of a completely new visual code, based on mathematics. Woody Vasulka: "The dramatic moment of the transformation into a binary code of energy events in time, as they may be derived from light, or the molecular communication of sound, or from a force field itself, gravity, or other physical initiation, has to be realized. The process of analog-to-digital and digital-to-analog conversion envelops the internal digital-code operations, the state of the world, which is exclusively man-organized and cross-disciplinary."[406] In 1976, the Vasulkas began to develop a new digital system together with Don MacArthur, who built a prototype, and Walter Wright, who wrote the first programs. Together with Jeffrey Schier, then still a student at New York State University, they built a more complex system, the so-called Digital Image Articulator, or Imager. This system could take two video inputs, digitize them, and then perform a series of operations on those two images based on logical functions derived from the arithmetical and logical unit. Depending on which logical function was operating, the numerical codes - and hence the images - were combined differently, but in absolutely predictable ways. Basically, one can describe this system as a mathematical digital system.[407] The process itself, the description of the screen in which each pixel or point is encoded to create the image, takes place on a syntactical level, rather than a semantic one. This linguistic analogue is nothing new, but was a good metaphor to describe this work. The mathematical 'language' based on logical functions created a unified language in which the single elements formed syntactical relationships which he called "syntax imaging." The criteria for the composition of the imagery were now determined through the articulation of the structural properties of ditigal space. The transformations and manipulations of the images only happen through the physical structure of the system. It was yet another step toward the development of a code, like syntax functions in language. The emphasis on inquiry has often been interpreted as being solely didactic. Yet, for Woody there is no doubt that "here and in video as well ... the hardware itself was a carrier of aesthetic definitions beyond my expectations," being within, and not coming from the outside.[408]

Steina: Machine Vision

Steina's work began to take a different turn during the time that they were in Buffalo, NY (1973-1979). Here she began her "Machine Vision" works, which bring to the fore how important this exchange between machine and (wo)man was for her. In *Allvision* (1976) she placed a bar on a turntable with two cameras mounted on either end with a mirrored globe in the middle. In each corner was a pair of monitors. Rotating the table 360 degrees, the cameras captured the whole room, including the observer, from a kind of fish-eye perspective. The camera sees both from behind, through the reflections in the sphere, and what is in front, although distorted and somewhat abstract. The movement gives the impression of a fluid environment. Other variations played with different positions of the mirror or sphere, with rotation and cyclical possibilities, zooming in and out, and so on. Steina wanted to explore situations in which the camera sees more than the human eye under normal conditions. Steina states: "The cameras alone scan the whole room. The idea was of course that the whole room can never be perceived or understood by human vision. Inserting the sphere in between emphasized the absurdity. When I mount the camera on the car, I define it as machine vision, but when I use the sphere, it is the concept of allvision."[409] Elsewhere she said: "These automatic motions simulated all possible camera movements without making the camera and its operator the center of the universe. Time and motion became the universe with its endless repetitive cycles and orbits."[410] There is no narrative with a beginning or end; there is only the eye of the camera registering the 'outside environment' as 'movement' in space and in time. The impressions of the images of the camera when turned 360 degrees, or of two cameras directed at each other, cause the registration of space to appear indistinct, yet multi-layered. Steina Vasulka's 'machine vision' explores the visions or rather the perceptual qualities of a machine (camera) in conditional situations. Or as she put it herself: "My Machine Vision installations were performing systems, and they occur in the studio, or out in the landscape or an exhibition..."[411]

Since their move to Santa Fe, New Mexico, in 1980, Steina has continued this approach and had consistently developed and adapted it in installations like *The West* (1983-1984), *Geomania* (1989) or *Borealis* (1993). The exploration of the phenomenology of space now includes the outdoors. She admits: "I moved there because I wanted to experience what it is to live in beauty. I did not want to think that it was going to affect my images as much as it did. For the first two years I resisted it. First of all because the beauty of the West is so seductive. And, secondly, I didn't feel up to it. ... I just went outside one morning and said 'Well, my studio doesn't have any walls and the ceiling is very high, and it's blue.' I just adopted the whole Southwest as my studio. So that's when I made peace with the idea that the landscape of the Southwest was going to be my image material."[412] Since the ordinary beauty of a landscape still remained suspect to her, she re-configured it into an imagery which appears to represent a metaphor for human intervention in nature at large. *The West* is a two-channel video installation about this landscape of the Southwest. It is a story of nature become landscape, of the cultivation of a piece of land once wild where "mankind leaving an imprint" is visible everywhere, through buildings, telephone cables, satellite and telescope systems (Los Alamos is near). Using the spherical mirror to create double vision and the horizontal drift to dissolve the visual frame of a motif, mixing natural and electronic colors, overlapping different images, she visualized the land as being in continuous motion, like a musical composition. The four-channel audio piece is fully integrated with the video images. In *Lilith*, a videotape made in 1987, she superimposed the face of a woman - painter and poet Doris Cross - on a landscape. Willingly or not, Steina thereby created a layering of metaphors for similar processes in all living beings on earth: a passage of time portrayed in the face of an elderly woman coincides with the changing faces of the land: mother earth. The footage for *Borealis* (1993) was filmed during a stay in Iceland, where she was born. On four large translucent screens suspended in 'space' close-up images of foaming and surging waters overwhelm the passing observer. The continuous coalescing movement and sound are more than a suggestion of the power and force of nature. There is an awe toward nature and the aesthetics of the natural environment, combined with a healthy scepsis towards false sentiment and sentimentality.

If one wants to distinguish between the work of the two artists, one can interpret Steina's recent work as a method to connect the machine vision with the reality of the external world, i.e. external from the point of view of the eye of the camera, whereas Woody's critical investigations of the function of the machine and its applications can be seen as a description of the internal workings of the machine, an exploration of the space within. They are investigating the same landscape as it were, but from different complementary perspectives, Woody from the inside and Steina from the outside.

Woody: Mapping Memories

The move to Santa Fe may also have influenced Woody. *The Commission* (1983), at least, appears a distinct deviation from his syntactical experiments. In *The Commission* Vasulka combines electronically generated imagery with a videotaped staged narrative. The story is based on the incident of a supposed commission given to Niccolo Paganini by Hector Berlioz, to compose a piece of music for him. It relates the last days of the legendary violinist - his fame and glory and the misfortunes surrounding his death and funeral. The 45-minutes-long videotape is divided into 11 segments. The opening section tells us that Paganini had lost his voice and had to speak through his "beloved illegitimate son." The son repeats what the father says, so that we may hear. Woody uses this element of repetition as a structural element throughout the videotape. In subsequent segments, Berlioz wanders aimlessly in a surreal landscape, contemplating Jungian notions of character, archetype and knowledge. The viewer experiences the death of Paganini and hears the whispered rumours, even before the man is buried. These filmic images are overlaid with computer-generated digitized veil-like wave forms. Woody Vasulka's 'opera' overlays, as it were, the visible camera obscura reality with the electronic reality. The current vision - presented as generated digitized imagery - overlays the past, or memory of the past - presented as filmed reality - which has come to us as a metaphor for the romantic myth of the artist as genius. Finally, the making of *The Commission* itself developed from an experiment into a metaphor for the rise of myths and legends surrounding fame and fortune - the process of art making and art politics. *Art of Memory* (1987) has been interpreted as the artist's battle with narration = photographic/filmic = chronological = linear versus non-narration = electronic = non-chronological = non-linear. Black-and-white photographs and film fragments of historic events of the first half of the century are superimposed on the electronically generated images of the landscape of the Southwest, the present as it were. The fragments are mainly war episodes: the Spanish civil war, the Russian Revolution, World War II and the atomic bomb. The fragments represent not only our own archival memory, jumbled together, as shreds of a narrative interrupted by forgotten parts; what is left to posterity of the past. They also reflect on how the construction of memory and history is mediated through the camera arts; on what we deem important to remember as truth; or may be not even that. Sturken interprets: "The incongruity of these images of history set against the dry forms of the American Southwest evokes a kind of timelessness; the desert landscape is emblematic of time marked within the earth, the past and future merged. *Art of Memory* is thus an attempt to situate the images of history within the fluid terrain of time, to mark their ephemerality."[413]

Contextualizing Virtual Space

The development in the works of the Vasulkas shows how the complexity of their research increased with the complexity of the technologies, how the philosophical and aesthetic implications became more and more multi-layered and interdisciplinary. Woody's exploration of inner space continued with the *Hybrid Automata*, which focused on the development of a robotic camera, with a "pan/tilt/rotate camerahead." The video camera is placed on a camera stand, with a 360 degree rotation axis, and connected to a computer, which is programmed so that it communicates with the environment on two different levels, a visual and an aural mode, 'acting/reacting' upon the optical and acoustic phenomena of the surrounding space. The 'synthetic' computer model is set up to confront the real acoustic and optical world, and to interpret the received data via learning capabilities in order to create a virtual environment. The robotic device, "when linked to graphic virtual computer space, attempts to redefine the straditional stage and established media space."[414] A number of questions are dealt with: Cartesian space and perspective are challenged by the parameters with which the computer driven mechanism maps the space; by doing so it gathers memory; the active engagement of the viewer is requested. The first stages of the project were presented during a*rs electronica* (1990) at Linz (Austria) as *The Theatre of Hybrid Automata* (1990). Woody Vasulka and David Dunn wrote: "This environment explores the potential for interpretation between diverse sensory worlds, subjective and objective experience, and the real versus the impossible through aural illusions of orbital motion in physical and virtual space."[415] Later on Woody wrote: "I see the *Theater of Hybrid Automata* as an apparatus that is conscious of space. The basic cyclical ritual of the machine, the process of calibration is a phenomenon performed at each and every moment whenever a machine charts its future. Perhaps it is complementary to our own human experience. From the moment of awakening, our mind begins the search for its identity, its alignment to time, the shape of the room, a street, a city. Finally, the way in which this assembly of objects, systems, and events behaves may help to trace some points of its original intent: to deconstruct, analyze, and describe the basic behavioral pattern of a techno-aesthetic system."[416]

Vasulka's research project is cross-disciplinary, involves work stations and what Gene Youngblood has called a "design team" (including composer/artist David Dunn, media writer/critic Gene Youngblood, poet Liz Rymland, and artist/writer Peter Weibel), to investigate both the technical and intellectual/philosophical questions. That this type of research is not just technical is self-evident, as it touches on concepts of computer interactivity, artificial intelligence and computer learning systems, connectionism (the science concerned with the functioning of neural networks in the brain and parallel computer processing) and such technological developments as virtual reality. Constituting a major expansion of the traditional concepts of system control, the significance of these aspects of digital space extends "beyond the domain of art to proffer an expansion of human imagination through the merger of artistic perception and scientific process."[417] His new project, called *The Brotherhood* (1995), consists of series of *Tables* built from scraps of industrial and military waste from nearby junkyards in Los Alamos and elsewhere, and investigates possibilities of a more fully interactive environment. Woody Vasulka about *Table III*: "In principle, the installation performs its preprogrammed audio/videocycle; the overall composition progresses as a single time line unless the audience intervenes. The visual concept is based on a single light beam, split and redirected to the six coordinates of a cube. At the walls of the imaginary cube, there are six translucent screens, showing the projected images on both sides of the screens."[418] It runs on an interactive laserdisc program. References and comments concerning the nuclear and space industries come to mind; thoughts about mankind's - mainly male, though - tendency to want to reorganize everything in nature; about man-machine relationships in general.

However, for the Vasulkas the concept of computer interactivity is understood to be more than the interface between the machine and the user/perceiver, hopefully opening up new potentials to describe the "real landscape," as a "landscape of synthesized images, which could create new sensory relationships - new perceptual experiences."[419]

The Artist as Mediator

The videoworks of the Vasulkas are 'steps' in a continuous, but not necessarily linear, research, and the viewer who steps into the middle perceives them as sections without an apparent beginning or end. Although the idea of process became a common element in the art of the sixties and seventies, it was generally concerned with the exposition of internal changes over a period of time with some kind of beginning or end. The Vasulkas' seemingly fragmentary segments were often too difficult to follow for the art world, accustomed to looking at a finished work of art, for the Vasulkas required placing the work in the context of another kind of process, that of an open unfinished research project. With a few exceptions, like artist Shalom Gorowitz, who described watching some of the early tapes as "an incredibly sensual experience," criticism was often restricted to situating the work of the Vasulkas in the realm of documenting research procedures.[420] On the one hand, this did not happen totally without the consent of the artists, who have denoted their procedures themselves as didactic video. On the other hand, the Vasulkas do perceive themselves as artists and place the purpose of their work in the realm of art. After all, it provides their economic base and gives them the freedom to experiment.

The relationship to science is their apperception of art as a problem solving activity, as a mode of inquiry, which may lead beyond the existing utilization of the hardware and software presented by the industries and to an evolution of the implied system aesthetics. Woody Vasulka has said about his position: "I come from a non-scientific discipline. But in going through the system I keep finding these coincidences, and then I try to rationalize them as I see them. I wouldn't be able to rationalize them beforehand as scientists do. They usually have a much clearer idea of what they're seeking. I've always been interested in ambiguity, or rather in magic, in imaging."[421] He has explained that he sees himself as a mediator of this kind of knowledge, making computer science available as a commonly utilized material with the general purpose to create works of art that contribute to the evolution of these technologies which they translate into 'personalized tools,' hoping that "the articulation of such syntactical principles will have direct application in such fields as scientific visualization and artificial reality research."[422] Interpreted this way, there is indeed an educational, didactic aspect implied in the research. Apart from that, there are several other levels of research to distinguish. First there is the research itself, as an explanatory mode of the internal functions of the new technologies. Secondly, by opening up the possibility of a personalization of these 'tools' or media, their working method literally transforms the notion of our accepting what the industries offer, with all the economic and political implications. Thirdly, they feel that artists "must help to shape what is quickly unfolding as a fundamentally new perceptual environment which is ushering forth profound epistemological changes...for the sake of cultural evolution and preservation."[423] So their longstanding interest in cognition and perception needs to be considered as well. How we receive images and retain them is just as important for the Vasulkas as it is to Bill Viola. Finally, the communication and information context of these media systems and how they affect human behavior in an automated environment are also part of the research.

The Vasulkas and their collaborators are convinced that the digital computer will bring totally new characteristics into sensory, linguistic and social communication, which will in turn have a direct impact on our mode of thinking. Woody Vasulka thinks that it will effect the way we look at reality, which he feels will express itself as a confrontation between the computer-generated imagery, which is an internally created reality, and external reality. But the discussion concerning the illusionary aspects of the artificial, synthetic image and the 'real' image will become less and less important.[424]

Summarizing, we can say that the Vasulkas' research of digital space leaves the boundaries of the art object in two directions: 1. externally, into an art that searches to bring together interdisciplinary functions, 2. internally, into an art that searches to merge accepted aesthetic conventions by creating a new perceptual mode of viewing the environment. We have already mentioned that Gene Youngblood coined the term "metadesign" to designate the work of artists who set out to create a new context rather than focus on content only. This "metadesigner" searches for new models, as it were, which may be technical, structural, perceptual, aesthetic and so forth, to be used by others, whether artists or other people, to create contexts. Accepting this definition, the Vasulkas would indeed fall into this 'category.'[425]

Conclusion

Design of Contexts

It is nothing new that those who hold the thread to the state-of-the-art technologies hold the economic power. So, the world will be - or perhaps already is - run by those who have access to and control the information technologies. This is exactly what the Vasulkas understood early on. The artist using advanced technology has to keep up with the rapid developments in the field. Artists like Robert Irwin or Alan Sonfist also soon understood that in order to work successfully in the public sphere they had to learn to speak 'the language' of developers and construction managers, city administrators and possible sponsors. This requires both a mentality and a devotion that not many possess.

For more than a decade Kit Galloway and Sherrie Rabinowitz (Mobile Image), two artists based in Los Angeles, have been searching and finding ways to get access to satellite and space technologies, among other things. In their opinion it is absolutely necessary to acquire the knowledge and learn about the structure with which this whole network of information technologies operates. Galloway and Rabinowitz even feel that it is essential for all of us to learn how the (tele)communication technologies function, like the relatively simple Internet software programs, for example, which means learning who and what is behind them, in order to be able to manage human and material resources properly. The artists' intention is to make us aware that the reality of the new mass communication networks no longer only consists of newspapers, radio, television, telephone or fax, but includes the total structure of invisible electronic and computer networks that have penetrated our whole environment. There is no doubt that the rapidly growing infrastructure of the communication and information technologies, including satellites, computer networks, datasystems and so on, is changing the economic world game drastically. One needs only glance at the electronic money transferral business, which is actually one computer talking to another over vast distances via cables, via satellites, very rapidly, 24 hours a day. Those with access to them will have the power and the wealth. The other field based on digital electronics is the entertainment industry, with its CDs, CD-Roms currently replacing the videocasette, DATs, etc. When he was still at the Massachusetts Institute of Technology futurist Peter

Schwartz, co-founder of the Global Business Network which advises the Pentagon and large corporations, prophesied that the whole financial structure would be dominated by this infrastructure, "linking things that are on the one hand a function of finance, the financial world, and on the other hand the electronic entertainment media."[426]

Mobile Image perceives its purpose to be the development of new models that may redefine the limits of these technologies for people's usage. So part of their research has consisted in looking at new hardware solutions, in designing a "cost-effective, kickass, multimedia, cross-cultural teleconference terminal that will allow communities of common concern to link up and evolve collectively."[427] The idea of connecting cross-cultural communities via this type of networking to familiarize them with the possibilities of telefax, full-motion video, slow-scan options, pictorial data management, or text conferencing found its first realization in a pilot project called the *Electronic Café*. It premiered at the 1984 Summer Olympics, in Los Angeles, as part of the Olympic Arts Festival. Mobile Image: "Electronic Cafe linked MOCA (Museum of Contemporary Art) and five ethnically diverse communities of Los Angeles through a state-of-the-art tele-communications computer database and dial-up bank designed as a cross-cultural, multi-lingual network of 'creative conversation.' From MOCA downtown, and the real cafes located in the Korean community, Hispanic community, black community, and beach communities of Los Angeles, people separated by distance could send and receive slow-scan images, draw or write together with an electronic writing tablet, print hard-copy pictures with the video printer, enter information or ideas in the computer datadase and retrieve it with Community Memory keyword search, and store or retrieve images on a videodisc recorder which held 20,000 images."[428] The artists have set up a permanent *Electronic Café* in Santa Monica from where they regularly organize conferences and events. They had, for example, set up a connection with Biosphere II in Arizona, and Steina Vasulka's *Studio X* in Santa Fe. By now there are several *Electronic Cafés* around the world, among others in Moscow. The premise of the artists to open up and create a new context for communication by introducing and making available network technologies considered beyond reach was rather advanced in the art world of 1984. In 1996 the mushrooming of the 'cyberspace' cafés with Internet connection makes it seem a normal communications café now, somewhat dated even. Yet the emphasis on human interaction still makes the project stand out. Galloway and Rabinowitz have a background in the visual arts, but they stress that they do not produce artifacts, that is objects, but living events. In doing so, Mobile Image calls the traditional function of the artist as object-maker into question. The real makers of the events are in fact the participants. It is the viewer's response that completes the work that is set up by the artists, and in the event defines the parameters of the system.

Jonathan Benthall once remarked that what the new technological media, such as photography, xerox, laser and holography, video and computer, had in common was that they were communication media, which by nature made them part of the larger communication system.[429] He meant that the utilization of the technologies as media in art should include not only their aesthetic features, but also involve their whole (public) structural organization. This is basically what Mobile Image does.

Gene Youngblood has claimed that 'models' such as those created by the Vasulkas or Mobile Image cannot be developed by traditional artists. That the 'new' artists will probably be both artist and technician/engineer, or artist and scientist. Indeed, if an artist wants to work creatively in the field of electronic media, whether video, computer or satellite technology, he or she will have to be knowledgeable in order to be able to convey the ideas to the engineer, but up to a certain level this expertise can still be that of an amateur. Youngblood reminded us of the "Renaissance amateur artist." I am not sure whether Youngblood's comparison has any real significance. Artists like Michelangelo or Leonardo were proficient in different areas, but they were exceptions. Actually most Renaissance artists were highly specialized. Professional specialization has certainly been a main feature of this century. On the other hand, in the nineties we have witnessed the rise of the amateur scientist-artist and the engineer-artist, indeed. Still, those who venture into the field of computer arts/graphics are predominantly artist-scientist-programmers who have access to the sophisticated state-of-the-art supercomputers which only the government, certain industries and a few universities can afford, such as IBM's Thomas J. Watson Research Center, Los Alamos, New Mexico, the Lawrence Livermore Laboratory, Livermore, California, or the Jet Propulsion Laboratory at the California Institute of Technology, Pasadena. Their acceptance by the art world is not likely, considering the consistent neglect of this type of work by the 'serious' art world. Rare are the artists who could call themselves equally familiar with art and electronic or computer technology, although there are some institutions which actively promote collaborative projects between scientists, engineers and artists. Ohio State University (Columbus, Ohio) and the New York Institute of Technology (New York) have set up interdisciplinary programs. The Advanced Computing Center for the Arts and Design (ACCAD) at Ohio State University, headed by computer graphics pioneer Charles Scuri, has developed an experimental program for artists and animators. The Massachusetts Institute of Technology has been singular with its interdisciplinary programs between the arts and sciences for over two decades. Nicolas Negroponte, director of the Media Laboratory, told Steward Brand: "The artists around the Lab have ideas about what to do with the technology, and the scientists have ideas about what to do with the arts. The impact of computers on the arts will be bringing out the artist in all of us. Much of it will be like hanging the child's painting on the icebox. It doesn't have any meaning outside the family circle, but it's very important to the local constituency. You'll see a return of the Sunday painter."[430] Some of the computer graphics that resulted from scientific computer research have become popular indeed. The Mandelbrot fractals, discovered by Benoit Mandelbrot, are widely admired for their beauty, although their discovery was accidental and for scientific purposes only. The same is true for the famous Jupiter "fly-bys," simulations of the Voyager I passing by Jupiter, which were created by James Blinn at the Jet Propulsion Laboratory.

Probably most media artists will remain amateur engineers or technicians, and only few will be educated in the two fields. However, this may change, as many people will no longer hold the same job all their lives. More important was the fact that in the recent development in the media arts and art in public places, many of the collaborations between artists and specialists in other disciplines were based on an exchange of ideas and knowledge. This has been a crucial element in opening up the gates of understanding of each other's language. It made possible a necessary change in attitude and therefore constituted the beginning of another road and role which these artists have pursued. It has subsequently affected the way in which the art works are produced, the work of art being the result of a collaborative effort, and no longer of the sole investigative journey in the studio.[431] The artists I discussed have all worked with scientists and engineers or technicians at different times and on different levels, such as the development of such tools as the Digital Image Processor of the Vasulkas, or the collaboration of Nancy Holt and James Turrell with astrophysics.

The fact that the artists discussed have presented (a part of) their work purposively outside of the realm of the museum and gallery in order to create a different context for art has advanced the search for a different function as well. Consequently, aesthetic decisions are made subject to environmental, architectural, landscape, and social factors, or the structure of television networks, or databases. There is no doubt that a chosen technology largely affects the final aesthetics of a work, it being part of its function. The complexity of the environmental, social and sometimes political issues dealt with, as well as the context in which they are placed or operate, has directed the emphasis more toward the quality and level of information and communication than the purely aesthetic aspects, or at least placed both on an equal footing. Nam June Paik has emphasized this in his writing. Thus instead of a formal and aesthetic approach, the artists try to create a work of art from a largely functional and operational perspective. It means that art is no longer made from the point of view that it is something autonomous and separate from society. This aspect too contributes to a possible new road which may look at art in a different way from the modernist tradition, a tradition which after all is only a few hundred years old.

The development of the artists we have discussed is characterized by a gradual change from a display of simple processes and systems towards works displaying a multi-layered complexity. Hans Haacke's early experiments with grass and wheat growing under artificial conditions, with climatic phenomena or hatching chickens were in fact simple displays of biological and natural systems. When his research came to revolve around complicated social and political issues that were analyzed systematically on a number of levels, the number of questions to be answered multiplied and Haacke's work became more and more layered. James Turrell's *Roden Crater* project is a prime example too. Alan Sonfist's relatively small experiments with natural systems have become large ecological landscape projects. And the Harrisons' research on the survival conditions of a crab, grew from indoor experiments to large outdoor interdisciplinary research projects and proposals for sustainable eco-systems. Their research has systematically taken into account multiple variables of a climatic, biological and human, social and economic nature. This development parallels the increasing complexity of the environmental situation itself and the accompanying technological advancement.

In the nineteenth and early twentieth century, scientific research still dealt with the handling of two-variable problems, such as the relationship between pressure and volume (gases), force to distance (objects), current versus voltage (electricity). The models in the social sciences show a parallel mode, analyzing such interdependencies as, capital and labor, supply and demand, balance of power versus trade. In the first half of this century scientific discoveries came to involve problems with a larger number of variables. Meanwhile the mathematization of science led to problem-solving in terms of numbers and tables. Problems of disorganized complexity as they were called by Daniel Bell, could now solved since the theories of probability and chance, i.e. statistics, advanced considerably. The most important characteristic of problem solving research after about 1940 is that new (computer) sciences are based on handling large numbers of variables, such as systems theory, cybernetics, information theory, game theory, etc. On a social and economic level, similar changes took place. Increasing complexity in the management of large-scale organizations, or systems have led to techniques that are able to handle multi-variable interacting aspects.[432]

Although there is no direct correlation between these developments in science and art, there is no doubt a similarity in the movement away from an art which perceives itself as part of a linear and objective progress toward a non-linear and subjective one. I have indicated the association with and continuation of the ideas of systems analysis and cybernetics. Although I did not want to label the road which these artists have followed since the early seventies, I feel attracted to the term suggested by artist and writer Roy Ascott: "field theory." Ascott writes. "Its terminology frequently employs ideas of transaction, interplay, net, web, reversibility, association, psychism, mulitple meaning and connectivity. ... It demands awareness of time, a reaching out to other disciplines and other operational modes of consciousness."[433]

The shift towards the replacement of the object by a non-objective situation that includes the surrounding space, has brought about another role for the spectator. Art has at different times in history been concerned with the role of the beholder, and has tried to include him or her in ways other than merely looking. Dada events, and the

Happenings and performances of the fifties, sought to push the observer with physical violence. In subsequent environmental pieces the visitor was often required to crawl into installations. The perceptual experience became a part of the meaning. Although I have not dealt with this aspect extensively, it seems likely that the scientific interest and research in the nature of perception at the time stimulated many artists. The participatory element has remained an important part of the environmental works. The new art in public places has continued in this direction, including the community in the preparatory stages of the process, even.

An important motivation for artists who began to experiment with video, believing that one could influence and change existing power structures of television, was the idea that it must be possible to create a possibility for the viewer to participate. However, the electronic and computer technologies have given the greatest impetus for a truly active participation, in that the viewer/participant can select a personal sequence of images and sound, albeit from an existing or pre-recorded database. If the production of a work of art in an interdisciplinary context affected the traditional hierarchy between artist and engineer/technician, the interactive computer network system has toppled the hierarchy between artist and viewer. Visionary Nam June Paik already flouted the hierarchical position of the artist, and so-called 'high art,' since the beginning of his career. Kit Galloway and Sherrie Rabinowitz (Mobile Image) also purposively put themselves in the background as authors of an art work.

When Walter Benjamin proclaimed the loss of the art work's aura in the age of mechanical reproduction, reproduction was the possibility to make many copies as similar as possible, from a unique original. In dataspace there may be a first version, but there is not even an original, no definite version, because each interruption in the stored information will create another status. An interactive computer work or a telecommunication work is in fact an open system and the result depends on the interactions of the participants. Some time ago already Bill Viola wrote: "On the video disc itself you're laying out programmed information. That means you can jump anywhere.... This is why the video disc is a newly emerging art form. It's the art of interactivity. One of the areas where we really need new skills is in deciding what our scheme for interactivity is. ... Soon images will be formed out of a system of logic, almost like a form of philosophy - a way of describing an object based on mathematical codes and principles rather than freezing its light waves in time."[434] In the meantime artists began to experiment with the interactive possibilities of the interactive videodisc and CD-Rom. Among the early works was Grahame Weinbred's *The Earlking*. These works have a pre-set structure in which a number of paths are delineated which the 'visitor' can follow. The 'player' creates his or her own work, but cannot add or alter its content. New interactive CD-Roms, such as The Residents' *Bad Day on the Midway* (1996), have become more complicated and leave ample room for the 'player's' own investigation and imagination.

The same year that Viola made this 'prophesy' to Raymond Bellour, WELL (Whole Earth 'Lectronic Link) was set up in the San Francisco Bay Area "as as computer conferencing system that enables people around the world to carry on public conversations and exchange private electronic mail."[435] WELL's roots are the same as those of the counterculture magazine *Whole Earth Catalog*, brainchild of Steward Brand two decades ago. After the immense success of the catalog - its successor *Whole Earth Review* is still in print - the initiators introduced *The Whole Earth Software Review*. Soon WELL was born, and Steward Brand became its first director. The history of the subsequent development of 'internetworking' from ARPANET, developed as a decentralized military, scientific and goverment communications system, to Internet - also the Net - is excellently described by Howard Rheingold. The important characteristic of a communication system like Internet is that it is not organized, and seems to function like chaos theory. It gives the 'surfer' direct on-line access to any information he or she wants and each day the amount of information avialable grows exponentially. Just as important a characteristic is its possibility to actively contribute and interfere.

It is obvious that by their very nature the digital technologies oppose the modernist concept of the art object, as well as the notions of authenticity and authorship. The moment when the spectator can actively interfere with the pre-set structure and include his own imagery and ideas, the discussion of authorship takes a seriously different turn: even if the original idea remains the product of the author/artist. Quite a few artists (over 5,000 according to Timothy Drucker in his anthology *Culture on the Brink: Ideologies of Technology*) have launched projects on Internet's World Wide Web, but it is still hard to tell which way it will go. One of the first works of art created for the World Wide Web was Antonio Muntadas's *The File Room* (http://fileroom.aaup. uic.edu/FILEROOM.html), which started as an installation at the Chicago Cultural Center in 1994. It is set up as a database archive for case histories of cultural censorship since the ancient Greeks. Writer Robert Atkins noted that the project started with some 450 entries and has grown tremendously as the public has added hundreds to the compilation.[436] *The File Room* work is a clear example of a work of art which has come into being through an (inter)active participation of the public, or even more it would not have existed without this intervention. Concepts of authorship or authenticity have no function here. Yet, in terms of the prevailing definition of art, the status of these interactive video/computer works is not at all clear. Artists working with interactive computer networks and Internet might completely have to review the concept of art.

Essentially, Art in Public Places and the communication Media Arts deal with problems of public space and the human body. The network technologies are shedding a new light on the perception and function of the public and private spheres. It is at this crossroad where both 'movements' come together, having witnessed the disappearance of the single art object. Also, both have developed toward an active participation of the public, albeit on different levels. However, in the end these works of art will remain faint reminiscences of the conceptual art expressions of the seventies, if they do not question the existing concept of art from the start. Maybe we will then see a change in the modernist concept of art itself happening at the end of this century; a true change in paradigm.

Notes

1 Quoted in Calvin Tomkins, "Onward and Upward with the Arts. Maybe a Quantum Leap," *The New Yorker*, 47, No. 51, February 5, 1972, p. 42. In an earlier interview Heizer had already referred to this aspect: "In the desert I can find that kind of unraped, peaceful, religious space that artists have always tried to put in their work. ... This very private, almost religious meditation of the artist before the work does give the new art a mystical quality that is part of the general effort to escape the commercial gallery-museum world." In: Roy Bongartz, "It's called Earth Art - And Boulderdash," *New York Times Magazine*, February 1, 1970, p. 16.

2 Robert Smithson, "Some Void Thoughts on Museums," *Arts Magazine*, February 1967. Reprinted in Robert Smithson, *The Writings of Robert Smithson*, edited by Nancy Holt, New York University Press, New York 1979, p. 58. Henceforth cited as *The Writings*.

3 William C. Lipke, "Earth Systems," *Earth Art*, Andrew Dickson White Museum of Art, Cornell University, Ithaca, NY 1969, n.p.

4 John Gruen, "Art Meets Technology," *World Journal Tribune, Magazine Section*, New York, October 2, 1966.

5 Douglas M. Davis, "Art and Technology - The New Combine," *Art in America*, January/February 1968, p. 28. Douglas Davis tried to get to some definition and context about the nature of the newly proclaimed relationship between art and technology and science. After a rather general survey of the artistic fields which had become involved in major technological innovations, Davis distinguished the use by artists of materials produced by recent technology, the use of tools and methods with the same derivation, the use of a new imagery suggested both by the tangible forms of technology, and the revelations of science, and the full partnership between artist and machine in a creative process. Only the last aspect presented something really new and radical in his opinion.

6 Jonas Mekas, "On New Directions, On Anti-Art, On the Old and New in Art," *Village Voice*, November 11, 1965. Reprinted in *Movie Journal*, Macmillan Co., New York 1972, p. 208.

7 Lucy R. Lippard & John Chandler, "The Dematerialization of Art," *Art International*, Volume XII/2, February 1968, pp. 31-36; Lucy R. Lippard, *The Dematerialization of the Art Object*, Praeger Publishers, New York 1973.

8 Lawrence Alloway, "Systemic Painting," *Systemic Painting*, Solomon R.Guggenheim Foundation, New York, 1966.

9 For example: "Today painting and sculpture are no longer wholly satisfying. We need an art of greater energy. We need an art of the total environment. We need an art that unites us with the real rhythms of our era. The art of light and movement is dynamic, environmental and inclusive. ... The old art was an object. The new art is a system." In: Willoughby Sharp, "Luminism and Kineticism," *Minimal Art*, New York 1968, pp. 316, 318.

10 Part of Norbert Wiener's research comprised the development of a scientific theory of messages. He saw the study of messages as "a means of controlling machinery and society, the development of computing machines and other such automata." Wiener defined "information as the name for the content of what is exchanged with the outer world as we adjust to it, and make our adjustment felt upon it." In: Norbert Wiener, *The Human Use of Human Beings*, Houghton Mifflin Company, Boston 1950. Cited from First Sphere Books Edition 1968, p. 19.

11 Quoted from "Systems of Cross-References in the Arts," by Lawrence Alloway, Mark Roskill and Nicolas Calas, *Arts Magazine*, Summer 1971, p. 16.

12 Michael Lingner, "Zur Konzeption künftiger öffentlicher Kunst," *Kunst im öffentlichen Raum*, ed. Volker Plagemann, Dumont Kunstverlag, Cologne 1989, p. 254. Quoted from: Niklas Luhmann, "Das Kunstwerk und die Selbstreproduktion der Kunst," *Delfin*, III, 1984, p. 57. Niklas Luhmann has developed a social theory using the methodology from general systems theory.

13 Michael Lingner, ibid., p. 255.

[14] See Rosalind Krauss, "Sculpture in the Expanded Field," *October*, 8, 1979, pp. 31-44; Lucy R.Lippard, *Overlay: Contemporary Art and the Art of Prehistory*, Pantheon Books, New York 1983; John Beardsley, *Earthworks and Beyond*, Abbeville Press, New York 1984; Jeffrey L.Cruikshank and Pam Korza, *Going Public: A field guide to developments in art in public places*, Arts Extension Service in cooperation with the Visual Arts Program of the National Endowment for the Arts, Amherst, MA 1988.

[15] *Culture in Action: A Public Art Program of Sculpture Chicago*, Seattle: Bay Press, 1985; *Mapping the Terrain: New Genre Public Art*, ed. Suzanne Lacy, Seattle: Bay Press, 1995; *But Is It Art? The Spirit of Art as Activism*, ed. Nina Felshin, Seattle: Bay Press, 1995.

[16] Wolfgang Preikschat, *Video: Poesie der Neuen Medien*, Weinheim 1987. Also: *Ars Electronica*, Linz, AU; *Leonardo*; SIGGRAPH's catalogues; *Mediamatic.*

[17] Jack Burnham, *Beyond Modern Sculpture - The Effects of Science and Technology on the Sculpture of this Century*, George Braziller, Inc., New York 1968, p. 12.

[18] Jack Burnham, *The Structure of Art*, George Braziller, Inc., New York 1970.

[19] Gail Scott, *A Report on the Art and Technology Program of the Los Angeles County Museum of Art,* ed. Maurice Tuchman, Los Angeles 1971, p. 125.

[20] Jan Butterfield, "Part I. The State of the Real: Robert Irwin discusses the Activities of an Extended Consciousness," *Arts Magazine*, June 1972, p. 48.

[21] Michael Heizer has only recently worked on a land-reclamation project, *Effigy Tumuli* (1990). Walter De Maria has mainly continued to exhibit in the museum and gallery context. I have thus decided not to allot these artists chapters of their own.

[22] The relationships between art and science, and art and technology reflect a complex history. One might describe the connections during the first half of the twentieth century as follows: 1. artists used new materials and techniques; 2. a small group of artists had knowledge of the scientific discoveries and used their concepts, as they saw fit, such as atomic models, field concepts, or 4th dimension time-space relationships; 3. those artists also discussed the possible similarities between art and science; 4. it was rare that an artist had an education in both disciplines. Naum Gabo was such an exception. See Linda Dalrymple Henderson, *The Fourth Dimension and Non-Euclidean Geometry in Modern Art*, Princeton University Press, Princeton, New Jersey 1983.

[23] Frank Popper wrote extensively on kinetic art. He also discussed the relationships between kinetic art and architecture, and environmental artforms. See Frank Popper, *Die Kinetische Kunst*, DuMont-Schauberg, Cologne 1975, and *Art - action and participation*, New York University Press, New York 1975.

[24] Susan Sontag, "One Culture and the New Sensibility," *Against Interpretation*, Farrar, Straus & Giroux, New York 1967. There was a flurry of English and American publications between 1963 and 1975, reflecting the different positions concerning the role of the arts in science and technology. A general bibliography was published in *Leonardo*: David. R. Topper and John H. Holloway, "Interrelationships between the Visual Arts, Science and Technology: A Bibliography," *Leonardo*, Vol. 13, 1980, pp. 29-33; David R. Topper and John H. Holloway, Interrelationships of the Arts, Sciences and Technology: A Bibliographic Up-date," *Leonardo*, Vol. 18, No .3, 1985, pp. 197-200.

[25] See Pulsa, "The City as Artwork," *Arts of the Environment*, ed. György kepes, George Braziller, Inc., New York 1972, pp. 208-221; On USCO, see Gene Youngblood, *Expanded Cinema*, E.P. Dutton & Co., Inc., New York 1970, pp. 347-348.

[26] John Perrault, "To Make It New," *Village Voice*, December 5, 1968, p. 16.

[27] In this book, I will focus on Kepes's theoretical writings, not his art.

[28] Judith Wechsler, "György Kepes," *György Kepes: MIT Years 1945-1977*, Hayden Gallery, April 28-June 9, 1978, Cambridge, MA 1978, p. 10.

[29] At the Massachusetts Institute of Technology, György Kepes met a situation which nowhere else existed. Among the scientists were Norbert Wiener (cybernetics), Harold Edgerton (physical sciences), Richard Held (psychology), Jacob Bronowski (mathematician, visiting professor in 1963), Arthur Loeb (chemical physics), Noam Chomsky (linguistics), Marvin Minsky (artificial intelligence), Jerome Wiesner (electronics).

[30] Stewart Brand, *The Media Lab. Inventing the Future at M.I.T.*, Penguin Books, Harmondsworth 1987, p. 134.

[31] György Kepes, ed., *The New Landscape in Art and Science*, 1967, p. 20. *The New Landscape in Art and Science* contains contributions by scholars from many different areas of expertise. There are contributions by sculptor Naum Gabo, physicist Bruno Rossi, art-historian Siegfried Giedion, architect Richard Neutra, psychologist Heinz Werner, cybernetician Norbert Wiener, and writer Charles Morris.

[32] Ibid., p. 18.

[33] György Kepes, ed., *The New Landscape in Art and Science*, Cambridge, MA 1956. Fourth printing, 1967, p. 260.

[34] Ibid., Introduction.

[35] Douglas Davis, *Art and the Future*, Praeger, New York 1973, pp. 115-119. Also: Charles Morris, "Man-Cosmos Symbols," *The New Landscape in Art and Science*, 1967, pp. 98-99.

[36] Kepes correctly noticed for example that "Quantity and measurement are no longer the central preoccupations of mathematics and science. The English scientist (he quotes) Jacob Bronowski has pointed out how problems of shape and relation now occupy the foreground of our organized thinking." In: György Kepes, *The New Landscape in Art and Science*, 1967, p. 173.

[37] "Interview: Billy Klüver," conducted by Suzanne Ramljek, *Sculpture*, May-June 1991, Volume 10, No. 3, p. 32.

[38] Douglas M. Davis, "Conversations with György Kepes, Billy Klüver and James Seawright," *Art in America*, January/February 1968, pp. 38-45.

[39] György Kepes, "Art and Ecological Consciousness," *Arts of the Environment*, ed. György Kepes, George Braziller Inc., New York 1972, pp. 1-12.

[40] György Kepes, "Toward Civic Art," *Leonardo*, Volume 4, No. 1, Winter/Hiver 1971, p. 72. Article based on a text published in connection with the exhibition *Explorations*, held at the Smithsonian Institution, Washington D.C., from 4 April-10 May, 1970.

[41] The previous significant period of federal support for public art forms was during the thirties through the Works Progress Administration (WPA) and Treasury Section Art Programs.

[42] György Kepes, "The Artists' Role in Environmental Self-regulation," *Arts of the Environment*, 1972, pp. 167, 171.

[43] Die Globale Revolution, Bericht des Club of Rome 1991, *Spiegel Spezial*, No. 1, 1991.

[44] *EAT News* (New York), Volume 1, No. 2, June 1, 1967, n.p.

[45] Billy Klüver, "The Pavilion," *Pavilion*, by Experiments in Art and Technology, ed. by Billy Klüver, Julie Martin and Barbara Rose, E.P. Dutton & Co., Inc., New York 1972, p. x; Billy Klüver, "Remarks by Billy Klüver," *EAT News*, Volume 1, No. 3 (November 1, 1967, n.p.); *TECHNE*, Volume 1, No. 2, New York, November 6, 1970, front page.

[46] See for the development of these ideas a 1977 lecture for the I(nternational) I(nstitute) of C(ommunication) Annual Conference 1977, Group Session V: Art and Technology. Lecture by Billy Klüver, "The Artists' Expertise For Communication Planning," published as *EAT Proceedings*, No. 13, December 15, 1977.

[47] Jack Burnham, *Beyond Modern Sculpture*, pp. 370, 373.

[48] Jack Burnham, "Notes on Art and Information Processing," *Software, Information Technology: Its New Meaning for Art*, Jewish Museum, New York 1970, pp. 10-14.

[49] The notion of a dissolution of art had been ventilated a little earlier in Jasia Reichardt's exhibition *Cybernetic Serendipity*. It was the first major exhibition of computer-based art works, ranging from programmed music scores, computer films, and computer graphics to cybernetic sculptures. See Jasia Reichardt, *Cybernetic Serendipity: The Computer and the Arts*, Institute of Contemporary Art, London 1968; published as a special issue of *Studio International*.

[50] Billy Klüver, "9 evenings: theatre and engineering," *Nine Evenings: Theatre and Engineering*, Foundation for the Performing Arts, New York, 1966, p. 3 (pages not numbered).

[51] John Gruen, "Art Meets Technology," *World Journal Tribune, Magazine Section*, New York, October 2, 1966. Quoted from *E.A.T. Clippings*, Volume 1, No. 1, (April 1966-July 1969), pp. 4, 5; interview with Robert Rauschenberg.

52 John Cage, statement in Douglas Davis, *Art and the Future*, 1973, p. 69 (photo caption). Originally from *Nine Evenings: Theatre and Engineering*, 1966, p. 4. David Tudor in Douglas Davis, *Art and the Future*, 1973, p. 70.

53 Nilo Lindgren, "Art and Technology II. A Call for Collaboration," 1969, *IEEE Spectrum*, May 1969, p. 50. Also: Nilo Lindgren, "Art and Technology I. Steps toward a new synergism," *IEEE Spectrum*, April 1969, pp. 59-68.

54 Ibid., p. 50.

55 John Gruen, "Art Meets Technology," *E.A.T. Clippings*, 1966, p. 4.

56 Brian O'Doherty, *Object and Ideas: An Art Critic's Journal*, Simon & Schuster, New York 1967. Quoted from Nilo Lindgren, "Art and Technology, II. A Call for Collaboration," p. 47.

57 Jack Burnham, "The Future of Responsive Systems in Art," revised version published in *Beyond Modern Sculpture*, p. 363.

58 Nilo Lindgren, "Art and Technology II. A Call for Collaboration," p. 49.

59 The organization of the *Nine Evenings* took enormous efforts and patience from all participants. "It is estimated that over 859 engineering hours went into the preparation , worth on paper at least $150,000. Nineteen engineers contributed more than 2,500 hours. The real cost of the extravaganza was more than $100,000. The total audience for the nine days was 10,000." See Douglas Davis, *Art and the Future*, p. 69; John Gruen, "Art Meets Technology," cited from *E.A.T. Clippings*, (April 1966-July 1969), pp. 4, 5. Published as *EAT Proceedings*, No. 7, July 1969, a compilation of 80 newspaper and magazine clippings, with a complete bibliography on E.A.T. until July 1969.

60 Nilo Lindgren, "Art and Technology II. A Call for Collaboration," pp. 47, 48.

61 One of the thorough art critiques was written by Lucy Lippard, but she discussed the *Nine Evenings* mainly within the context of the visual arts. For Lippard, the result was the basis upon which to judge a work of art, not the processes behind it, or technical failures. See Lucy R. Lippard, "Total Theatre," *Art International*, January 20, 1967, p. 39. See also: Simone Whitman, "Theater and Engineering - An Experiment," *Artforum*, February 1967, pp. 26-33.

62 Douglas Davis, *Art and The Future*, pp. 68, 72.

63 György Kepes, "The Visual Arts and the Sciences: A Proposal for Collaboration," *Daedalus*, 94, No. 1, 1965, pp. 117-133.

64 Billy Klüver, "The Artists' Expertise for Communal Planning," I.I.C Annual Conference 1977, Group Session V: Art and Technology, September 1977, *EAT Proceedings*, No. 13, December 1977, n.p.

65 Suzanne Ramljek, "Interview: Billy Klüver," p. 34.

66 From the E.A.T. "manifesto," signed by Billy Klüver and Robert Rauschenberg, *EAT News*, Volume 1, No. 2 (June 1, 1967), front page. Experiments in Art and Technology published a regular newsletter, *EAT News*, to keep its membership informed about the activities. After the fifth issue, which was published in March 1968 (edition 10,000), E.A.T. decided to change formats. From then on a bulletin, entitled *E.A.T. Operations and Information*, appeared several times a year, listing and describing its services. It was followed by *TECHNE: A Projects and Process Paper,* which was to present "projects, processes, and physical problems and the opportunities presented by the relationship between rapidly expanding technology and the individual." All completed projects, lectures, talks, listings etc. were well documented and reported in what were called the *Proceedings* of E.A.T. The *Proceedings* describe the initial, fast growth of the organization between 1967 and 1969. For a general chronological survey, see Marga Bijvoet, "How Intimate Can Art and Technology really be? - A Survey of the Art and Technology Movement," *Culture, Technology and Creativity in the late twentieth century*, ed. by Philip Hayward, John Libbey and Company, Ltd., London, Paris, Rome 1991, pp. 15-37.

67 Documented in *E.A.T.Proceedings*, No. 9, May 19, 1969.

68 *E.A.T. Proceedings*, No. 9, May 1969, p. 7; *EAT News,* Volume 1, No. 3, November 1, 1967, listing of Board of Directors and Council of Agents.

69 Maurice Tuchman, ed., *A Report on the Art and Technology Program*, Los Angeles County Museum of Art, Los Angeles 1971, p. 9. Writer Jane Livingston evaluated the project and expressed her doubt about longer-lasting future collaborations: Jane Livingston, "Thoughts on Art and Technology," *A Report,* pp. 43-47. However, the report relates that those artists who

were able to strike up a good personal relationship with the collaborating engineer or scientist were happiest about the whole project, where the interdisciplinary collaboration, the production as a team effort, became part of the total experience of the work.

70 Billy Klüver, "Remarks by Billy Klüver," *EAT News*, Volume 1, No. 3, (November 1, 1967), n.p.

71 See *TECHNE*, Vol. 1, No. 2, November 6, 1970, front page, and EAT Newsletters.

72 When E.A.T. was approached by Pontus Hultén with a request to provide works for his exhibition *The Machine as Seen at the End of the Mechanical Age* (Museum of Modern Art, New York, 1968), it offered to sponsor a competition for the most inventive technical contribution by an engineer to a work of art produced in collaboration with an artist. The competition was announced on November 12, 1967 in the New York Times, Scientific American and fifteen technical journals. Hultén chose six for his exhibition, all the other works proposed were eventually exhibited at the Brooklyn Museum as *Some More Beginnings* and received remarkable coverage from the newspapers, television and magazines. See *TECHNE,* Vol. 1, No. 1, 1969, front page; *Some More Beginnings,* Brooklyn Museum of Art, Brooklyn, New York, 1969, and *E.A.T. Proceedings,* No. 6, 1969.

73 The India project is briefly discussed by Robert Rauschenberg in Douglas Davis, "Robert Rauschenberg: Technology as Nature," Art and the Future, Thames & Hudson, Praeger, London, New York 1973, pp. 142-143 Also in: Billy Klüver, "The Artists' Expertise for Communication Planning," lecture presented at the I(nternational) I(nstitute) of C(ommunication) Annual Conference 1977, Group Session V: Art and Technology. Published as *E.A.T. Proceedings*, No. 13, December 15, 1977.

74 Douglas Davis, "Robert Rauschenberg: Technology as Nature," *Art and the Future*, p. 144.

75 For a background history of the Pavilion: Billy Klüver, "The Pavilion," *Pavilion*, by Experiments in Art and Technology, ed. Billy Klüver, Julie Martin and Barbara Rose, E.P. Dutton & Co., Inc., New York 1972, p. ix (pp. ix-xvi) Also: *TECHNE,* Vol. 1, No. 2, November 6, 1970, front page, p. 2. There was a little booklet which did not seem to do justice to the expenditure that went into this extravaganza and supposedly prestigious project., published as *E.A.T.Proceedings,* No. 8, March 25, 1970.

76 Nilo Lindgren, "Into the Collaboration," *Pavilion*, p. 10.

77 In addition, there was a program of performances, music and dance. Participating performers included now well-known artists: dancer Remy Charlip, performance artist Alan Kaprow, composer-musicians Alvin Lucier, Terry Riley, Pauline Oliveros, David Tudor, La Monte Young and Marion Zazeela.

78 There was already an interdisciplinary group of artists, scientists, writers , who were interested in the possible relationships between art and technology, meeting in Los Angeles before Experiments in Art and Technology was founded, and who gathered on a regular basis. This is not mentioned anywhere, neither in the *Pavilion* book, nor in the catalogue of the *Art and Technology Program*, and deserves further attention. I am grateful to Joan Hugo, who also participated in this program, for bringing this to my attention.

79 Jack Burnham, "Art and Technology: The Panacea that Failed," *Video Culture*, ed. John G. Hanhardt, Visual Studies Workshop, Rochester, NY 1986, pp. 232-248. From Calvin Tomkins we learn that Donald M. Kendall, President of PepsiCo Inc., was actually persuaded to put E.A.T. in charge of the Pepsi-Cola Pavilion at Expo '70. "Klüver and his colleagues were ejected from the premises, and E.A.T.'s relations with big business went into a precipitous decline." In: Calvin Tomkins, "Outside Art," *Pavilion*, pp. 105-165.

80 Billy Klüver, "The Pavilion," *Pavilion*, p. ix.

81 Ibid., p. xv.

82 Calvin Tomkins, "Outside Art," *Pavilion*, p. 162.

83 Jack Burnham, "Art and Technology: The Panacea that Failed," *Video Culture*, p. 236.

84 Selected articles about the CAVS and its activities: Jane H. Kay, "Art and Science on the Charles," *Art in America*, Summer 1967, pp. 62-67; Jud Yalkut, "Conversations with György Kepes," *Arts Magazine*, May 1970, p. 16-18.

85 György Kepes, "The Lost Pageantry of Nature," *Artscanada*, December 1968, p. 32.

86 Douglas M. Davis, "Conversations with György Kepes, Billy Klüver and James Seawright," *Art in America*, January/February 1968, p. 38-40.

87 György Kepes, "The Lost Pageantry in Nature," p. 33.

[88] See fn.12 in *György Kepes MIT Years, 1947-1977.*

[89] György Kepes, "The Lost Pageantry in Nature," pp. 31-32.

[90] Douglas Davis, "Conversations with György Kepes, Billy Klüver and James Seawright," 1968, p. 39. See *Boston Celebrations, Jubilee Projects by the Fellows of the Center for Advanced Visual Studies at MIT*, March 18-April 29, Institute of Contemporary Art, Boston 1975, p. 34.

[91] For a listing of the proposals see *Boston Harbor Project, 1968-70,* CAVS/MIT Press, Cambridge, MA 1968.

[92] Jonathan Benthall, "Kepes Center at MIT," *Art International*, Vol. XIX/1 January 1975, pp. 28-49.

[93] Douglas Davis, *Art and the Future*, p. 115.

[94] *Multiple Interaction Team*, CAVS/MIT Press, Cambridge, MA 1972, n.p.

[95] Most of the works had been conceived for the Sao Paulo Biennial, in 1969. The American artists withdrew their contributions at the last moment in support of the protest of their Brazilian colleagues against the militaristic government then. In: Jud Yalkut, "Sao Paulo, No! Smithsonian, Yes!," *Arts Magazine*, May 1970, pp. 16-18.

[96] In fact, Lowry Burgess' project went through various stages of execution, which he explained later on. In: Lowry Burgess, "Art and Science Project: 'Waiting-for-Light-Planes'," *Leonardo*, Vol. 7, No. 4, 1974, pp. 329-331.

[97] *ARTTRANSITION*, (October 15-19) University Film Study Center, CAVS/MIT Press, Cambridge, MA 1975. In the introduction, "Arttransition," Piene laid down his ideas about the future directions of the CAVS.

[98] György Kepes, "The Lost Pageantry of Nature," pp. 33-39.

[99] Recently the Center for Advanced Visual Studies has presented a survey of its activities since its inception in Karlruhe, Germany. For a full listing of CAVS exhibitions, see *Otto Piene und das CAVS*, Badischer Kunstverein, Karlsruhe 1988.

[100] Jonathan Benthall, "Kepes Center at MIT," pp. 30, 31.

[101] Emily Wasserman, "Explorations - toward a civic art," *Artforum*, June 1970, pp. 87-88 Also: György Kepes, "Toward Civic Art," *Explorations,* Smithsonian Institution, National Gallery of Fine Arts, Washington D.C. 1970, n.p.

[102] Jack Burnham, "Art and Technology: The Panacea that Failed," *Video Culture*, p. 241.

[103] Emily Wasserman, "Explorations - toward a civic art," pp. 87-88.

[104] The only writer who seriously wrote about the "artist as ecologist" in the early seventies was Gene Youngblood. Although such topics like environmental design, environmental landscape gardening and such were much discussed in the respective disciplines and professional magazines, there was no mention of them in the major art publications. Gene Youngblood, "World Game: the artist as ecologist," *Artscanada*, August 1970, pp. 42-49 and "Buckminster Fuller's World Game," *Whole Earth Catalog,* March 1970.

[105] Cynthia Goodman, *Digital Visions*, Harry N. Abrams Publishers, New York, Everson Museum of Art, Syracuse 1987, pp. 38, 43. Interview with Billy Klüver.

[106] Frank Popper, *Art - action and participation*, New York University Press, New York 1975, pp. 210-211.

[107] Tom Stonier, *Information and the Internal Structure of the Universe*, Springer Verlag, London 1990.

[108] Steward Brand, *The Media Lab. Inventing the Future at MIT*, Penguin Publishers, Harmondsworth, 1987, p. 79; Gregory Bateson, Steps to an Ecology of Mind, Ballantine, New York 1972.

[109] Steward Brand, *The Media Lab: Inventing the Future at MIT*, pp. 150-151.

[110] Barbara Rose, "Art as Experiment, Environment and Process," *Pavilion*, p. 104.

[111] Gene Youngblood, "The Open Empire," *Studio International*, April 1970, pp. 177-178.

[112] Jane Livingston, "Thoughts on Art and Technology," *A Report*, p. 47.

[113] Burnham adopted the still prevalent theoretical ideas of Shannon & Weaver's (mathematical) Information Theory concerning the measurement of the amount of information contained in a message: "The objective of Software is to stress the fact that information is simply a measure of response between sender and receiver," and connected them with ideas about art as an information system: art as message. See Jack Burnham, "Notes on Art and Information

Processing," *Software, Information Technology: Its New Meaning for Art*, Jewish Museum, New York 1970, p. 13.

[114] Bitite Vinklers, "Art and Information," *Arts Magazine*, September-October 1970, p. 47, 49. Also: Thomas B. Hess, "Gerbils ex Machina," *Art News*, December 1970, p. 73.

[115] Among the 'conceptual/idea' artists invited were Vito Acconci, David Antin, Lawrence Wiener, John Baldessari, Douglas Huebler, Joseph Kosuth, Robert Barry, John Giorno, and Allan Kaprow.

[116] Hans Haacke, in *Software, Information Technology: Its New meaning For Art*, p. 34. Due to technical failures Haacke was not able to execute the work. It was not until Documenta 5, in 1972, that he finally saw the piece realized.

[117] Theodor Nelson/Ned Woodman, "Labyrinth: An Interactive Catalogue," ibid., p. 18.

[118] Nicolas Negroponte, "The Architecture Machine Group, MIT," ibid., p. 23.

[119] Jonathan Benthall, *Science and Technology in Art Today*, Thames & Hudson, London 1972, p. 84.

[120] The organizing committee wanted the exhibition to move beyond the *Cybernetic Serendipity* exhibition (ICA, London 1969), to address in part the changes in "personal and social sensibilities altered by this revolution," and it was decided to start from the notion of software. The term software was extensively defined in the catalogue, but its application was heavily criticized even before the show opened. In: Robert Mallory, "Notes on Jack Burnham's Concepts of a Software Exhibition," *Leonardo*, Vol. 3, No. 2, 1970, pp. 189-190.

[121] For the small list of articles in art periodicals, published in the late sixties/early seventies, see Cynthia Goodman, *Digital Visions*, 1987.

[122] Albert Einstein, *The World as I See It*, New York 1934, p. 65.

[123] Stephen W. Hawking, *A Brief History of Time*, Bantam Books, New York 1988, pp. 55-56.

[124] Erwin Schrödinger tried to visualize the paradox implied in Heisenberg's Uncertainty Principle with a thought experiment that became known as "Schrödinger's Cat."

[125] John Gribbin, *In Search of Schrödinger's Cat. Quantum Physics and Reality*, Wildwood House, London 1984. Corgi Books, pb. edition, pp. 151-152.

[126] Ludwig von Bertalanffy, "The History and Development of General System Theory," *Perspectives on General System Theory*, George Braziller, Inc., New York 1974.

[127] Ludwig von Bertalanffy, *General System Theory*, George Braziller, Inc., New York 1968, p. 37, and: "The History and Development of General System Theory," *Perspectives on General System Theory*, 1974, p. 154.

[128] Thomas Kuhn had defined the scientific developments of the twentieth century as a major shift in paradigm. "If science is the constellation of facts, theories, and methods collected in current texts, then scientists are the men who, successfully or not, have striven to contribute one or another element to that particular constellation". In: T.S.Kuhn, *The Structure of Scientific Revolutions*, University of Chicago Press, Chicago 1970, p. 1.

[129] Max Planck, *Where is Life Going?*, London 1933, p. 24.

[130] Jack Burnham, "The Aesthetics of Intelligent Systems," *On the Future*, Viking Press, New York 1970, p. 111.

[131] Jack Burnham, *Beyond Modern Sculpture*, p. 11.

[132] Jonathan Benthall criticized: "To read Burnham on vitalist or biomorphic sculpture is to be reminded that there is no apparent correlation between the stature of a given artist and the validity of his scientific assumptions. The discontinuities between science and modern art are as interesting as their interaction." In: Jonathan Benthall, "The prehistory of the invisible," *Studio International*, November 1969, p. 152.

[133] Jack Burnham, *Beyond Modern Sculpture*, p. 369.

[134] Ibid., p. 369.

[135] Because "by this method scientific findings are used instrumentally, to construct a conceptual framework adequate to the understanding of nature as an integral network of ordered interdependency of which man is a part. ... The universe here conceptualized is an emergent-holistic universe, with systemic processes supervening over the mechanical processes and

resulting in the local reversal of the generally entropic trends in the cosmos," said Ervin Laszlo, one of the pioneers in systems thinking in philosophy in 1972. In: Ervin Laszlo, *Introduction to Systems Philosophy*, Gordon and Breach, New York, London, Paris 1972, p. 12.

136 Michael Fried, "Art and Objecthood," *Artforum*, Summer 1967, pp. 12-23.

137 Jack Burnham, "Systems Esthetics," *Artforum*, September 1968, p. 32.

138 Ibid., p. 34. Also: Jack Burnham, "Real Time Systems," *Artforum*, September 1969, pp. 49-55.

139 Ibid., pp. 32, 35.

140 Jack Burnham, "Notes on Art and Information Processing," *Software, Information Technology: Its New Meaning for Art*, pp. 10-14.

141 Jack Burnham, "The Aesthetics of Intelligent Systems," *On the Future*, p. 121.

142 Jack Burnham, "Art and technology: The Panacea That Failed," *Video Culture*, p. 245.

143 Robert Smithson, "Quasi-Infinities and the Waning of Space," *Arts Magazine*, November 1966, pp. 28-31.

144 Paul Ryan: "Cybernetic guerilla warfare... because the tool of portable video is a cybernetic extension of man and because cybernetics is the only language of intelligence and power that is ecologically viable." It is concerned with asking the right questions about "how does power function, ... how is this system of communication maintained," what information is habitually withheld and how," etc. In: Paul Ryan, "Cybernetic Guerilla Warfare," *Radical Software*, No. 3, Spring 1971, pp. 1-2.

145 Paul Ryan, "Cable Television: The Raw and Overcooked," *Radical Software*, No. 1, Summer 1970, p. 12.

146 Nam June Paik, "Norbert Wiener and Marshall McLuhan," *Institute of Contemporary Arts Bulletin*, London 1967. Reprinted in: Nam June Paik, *Videa 'n' Videology*, Everson Museum of Art, Syracuse, NY 1974, pp. 27-29 and pp. 24-25.

147 Peter Drucker, *The Age of Discontinuity*, Heinemann, London 1969, p. 250.

148 Michael Shamberg, *Radical Software*, No. 6, 1971, p. 1.

149 *Radical Software*, No. 2, cover.

150 Michael Shamberg, "Guerilla Television - Media America," *Radical Software*, No. 6, 1971, p. 13.

151 Michael Shamberg, "Guerilla Television - Media America," *Radical Software*, No. 6, 1971, p. 26.

152 Howard Rheingold, *The Virtual Community*, Minerva, London 1995.

153 Leonardo, Vol. 1, No. 1, editorial page.

154 Leonardo, Vol. 16, No. 1, 1983, p. i.

155 Leonardo, Vol. 16, No. 2, 1983, pp. i-iii.

156 Jack Burnham, "Hans Haacke: Wind and Water Sculpture," *Tri-Quarterly Supplement*, Number 1, Northwestern University Press, Evanston, Spring 1967, pp. 1, 7.

157 Jeanne Siegel, "An Interview with Hans Haacke," *Arts Magazine*, May 1971, p. 18.

158 For an extensive description of his "systems method," see Edward F. Fry, *Hans Haacke Werkmonographie*, DuMont Kunst-Praxis, Cologne 1972, pp. 32-35.

159 Hans Haacke, statement in *Earth Art*, Andrew Dickson White Museum of Art, Cornell University, Ithaca, NY 1969, n.p.

160 Jack Burnham, "Systems Esthetics," p. 35.

161 Jeanne Siegel, "An Interview with Hans Haacke," p. 18. Also: John Noel Chandler, "Hans Haacke: The Continuity of Change," *Artscanada*, June 1969, pp. 8-11.

162 For an explanatory description of ZERO's intentions, see Otto Piene & Heinz Mack, *ZERO*, Dumont-Schauberg, Cologne 1958. (English translation by H.Beckman, 1973)

163 Jeanne Siegel, "An Interview with Hans Haacke," p. 19.

164 Ibid., p. 18.

165 Howard Wise Gallery in New York actively supported kinetic art forms in the sixties. Wise was also among the first to register the transformations in technology-based art forms. Already in 1965 he showed computer graphic works, and in 1969 he premiered the first works made with video, *TV As A Creative Medium* (New York 1969).

166 For Haacke's documentation of the project see Edward F. Fry, *Hans Haacke: Werkmonographie*, pp. 55-70.

[167] Hans Haacke, "Visitors' profile: a statistical breakdown of spectators," *Software, Information Technology: Its New meaning For Art*, p. 34.

[168] All quotes are from Howard S. Becker, & John Walton, "Social Science and the Work of Hans Haacke," *Framing and Being Framed. 7 Works 70-75*, Nova Scotia College of Art and Design, Halifax 1975, pp. 145-152. One later article on the subject appeared in Kunstforum: Astrid Wick-Kmoch, "Kunst + Systemtheorie + Sozialwissenschaften: Zu den Arbeiten von Hans Haacke," *Kunstforum International*, Bd.27 (1978), pp. 125-142.

[169] Benjamin H.D. Buchloh, "Hans Haacke: Memory and Instrumental Reason," *Art in America*, February 1988, pp. 97-107, pp. 157-159; Fredric Jameson, "Hans Haacke and the Cultural Logic of Postmodernism," *Hans Haacke: Unfinished Business*, Museum of Contemporary Art, New York & MIT Press, Cambridge, MA 1986, pp. 42-43; Yve-Alain Bois, Douglas Crimp and Rosalind Krauss, "A Conversation with Hans Haacke," *October*, 30, Fall 1984, pp. 49-81.

[170] Hans Haacke, "Museums, Managers and Consciousness," *Art museums and big business*, edited by Ian North, Art Museums Association of Australia, Kingston 1984, pp. 33-40, reprinted in *Hans Haacke: Unfinished Business*, Museum of Contemporary Art, New York & MIT Press, Cambridge, MA 1986, pp. 60-73.

[171] Quoted from a text written by Hans Haacke for the catalof of the 45th Biennale di Venezia, Venice 1993, p. 173.

[172] Robert Smithson, *The Writings of Robert Smithson*, edited by Nancy Holt, New York University Press, New York 1979, henceforth cited *The Writings*. Also: Roy Bongartz, "It's called Earth Art and Boulderdash," *New York Times Magazine*, February 1, 1970, p .16; John Perrault, "Long Live Earth," *Village Voice*, October 17, 1968, p. 17; John Perrault, "Down to Earth," *Village Voice*, February 13, 1969, p. 18; John Perrault, "Nonsites in the News," *Village Voice*, New York, February 24, 1969, pp. 44-46.

[173] Lawrence Alloway, "Robert Smithson's Development," *Artforum*, November 1972, p. 54. A few writers have related Robert Smithson's work to a systems approach, possibly because the artist never referred to his work in this manner and also because he actually criticized Jack Burnham's definitions of a système d'art. See Anthony Robbin, "Smithson's Non-site Sights," *Art News*, February 1969, pp. 50-53.

[174] Robert Hobbs, *Robert Smithson: Sculpture*, Cornell University Press, Ithaca & London 1981.

[175] Robert Smithson: "Some Void Thoughts on Museums," *Arts Magazine*, February 1967. Smithson: "What is needed is an esthetic method that brings together anthropology and linguistics in terms of 'buildings.' This would put an end to 'art-history' as sole criterion. Art at the present is confined by a dated notion, namely 'art as criticism of earlier art.'" In: Robert Smithson, "Towards the Development of An Air Terminal Site," *Artforum*, Summer Issue, June 1967, p. 35.

[176] Smithson referred to linguist/philosopher Alfred Ayer. "Recently, there has been an attempt to formulate an analog between 'communication theory' and the ideas of physics in terms of entropy. As A.J. Ayer has pointed out, not only do we communicate what is true, but also what is false. Often the false has a greater 'reality' than the true. Therefore it seems that all information, and that includes anything that is visible, has its entropic side. Falseness, as an ultimate, is inextricably a part of entropy, and this falseness is devoid of moral implications." In: Robert Smithson, "Entropy and the New Monuments," *Artforum*, June 1966, p. 29.

[177] Claude E. Shannon, "A Mathematical Theory of Information," *Bell System Technical Journal*, No. 27, 1948, pp. 379-432, pp. 632-656.

[178] Quoted from "Entropy Made Visible," Interview with Alison Sky, *On Site* #4, 1973. Reprinted in *The Writings*, p. 190.

[179] George Kubler, *The Shape of Time: Remarks on the History of Things*, Yale University Press, New Haven, CT 1962.

[180] Robert Smithson was an avaricious reader, with multiple interests, as his references and the collection of books in his library show. There is an unpublished manuscript of Robert Smithson's library, c. 1974, compiled by writer/critic Valentine Tatransky, on file at the Herbert F. Johnson Museum of Art, Cornell University, Ithaca, NY.

[181] Paul Cummings, "Interview with Robert Smithson for the Archives of American Art/Smithsonian Institute," *The Writings*, p. 156.

[182] Quoted from: "..The Earth, subject to Cataclysms, is a Cruel Master," Interview with Gregoire Mueller, *Arts Magazine*, November 1972, p. 181.

[183] Robert Smithson, "Entropy and the New Monuments," *Artforum*, June 1966, p. 26.

[184] Physicist Erwin Schrödinger had already considered the statistical meaning of entropy, in his book *What is Life?* (1944), following upon Ludwig Boltzmann's investigations. Ludwig Boltzmann described the processes during the melting of crystals: [you] "destroy the neat and permanent arrangement of the atoms and molecules and turn the crystal lattice into a continually changing random distribution." In: Tom Stonier, *Information and the Internal Structure of the Universe*, Springer Verlag, New York 1990, p. 37.

[185] "I was still working with the resolution of the organic and the crystalline, and that seemed resolved in dialectics for me. And so I created the dialectic of site and nonsite. The nonsite exists as a kind of deep three-dimensional map that points to a specific site of the earth. And that's designated by a kind of mapping procedure." In: Paul Cummings, "Interview with Robert Smithson for the Archives of American Art/Smithsonian Institution," *The Writings*, p. 155.

[186] Robert Hobbs, *Robert Smithson: Sculpture*, p. 62. The artist discussed his ideas about crystallography extensively for the first time in an article about Donald Judd. "Donald Judd," in *7 Sculptors*, Institute of Contemporary Art, University of Pennsylvania, Philadelphia 1965. Reprinted in *The Writings*, p. 22.

[187] William C. Lipke, "Fragments of a Conversation," *The Writings*, pp. 168-170.

[188] See also Robert Hobbs, *Robert Smithson: Sculpture*, p. 79.

[189] Robert Smithson had in his library J.E. Cirlot's Dictionary of Symbols, in which all the symbolic meanings of the spiral are described.

[190] Smithson made a number of works in which he questioned the Renaissance perspective as an absolute truth in depicting space. The *Enantiomorphic Chambers* (1965) were two steel structures which were to hold mirrors placed at oblique angles, so that the viewer standing in front of the mirrors would see reflections of reflections. "This negates any central vanishing point, and takes one physically to the other side of the double mirrors. It is as though one were being imprisoned by the actual structure of two alien eyes." In: Robert Smithson, "Interpolation of the Enantiomorphic Chambers," *The Writings*, p. 39.

[191] On the *Spiral Jetty* see Robert Smithson, "The Spiral Jetty," *The Writings*, pp. 109-116.

[192] Robert Smithson, "Frederick Law Olmsted and the Dialectical Landscape," *Artforum*, February 1973, p. 65.

[193] Robert Smithson, "Toward the Development of An Air Terminal Site," *Artforum*, Summer Issue 1967, p. 37.

[194] "My own experience is that the best sites for 'earth art' are sites that have been disrupted by industry, reckless urbanization, or nature's own devastation. For instance the Spiral Jetty is built in a dead sea, and The Broken Circle and Spiral Hill (NL-Emmen 1970) in a working quarry." In: Robert Smithson, "Frederick Law Olmsted and the Dialectical Landscape," 1973 Reprinted in *The Writings*, p. 119, 124.

[195] Robert Smithson, *The Writings*, pp. 220-221.

[196] Among the corporations he approáched were Hanna Coal Company, Ohio Reclamation Association (Ohio), Peabody Coal, Bingham Copper Mine (Utah), Minerals Engineering Company (Colorado).

[197] Robert Hobbs, *Robert Smithson: Sculpture*, p. 219.

[198] Calvin Tomkins, "Onward and Upward with the Arts," *The New Yorker*, 47, No. 51, February 5, 1972, p. 55.

[199] Robert Morris, *Sonsbeek 71*, Arnhem, Netherlands, Part 2, pp. 57-59.

[200] See Lucy R. Lippard, *Overlay: Contemporary Art and the Art of Prehistory*, Pantheon Books, New York 1983; John Beardsley, *Probing the Earth: Contemporary Land Projects*, Hirshhorn Museum and Sculpture Garden, Smithsonian Institution, Washington D.C. 1977; pp. 32-35 and

pp. 66-70; Marga J. Bijvoet, "In Search of Self through Nature: Art Experiments in the California Deserts," *Arts and Architecture*, Winter 1983, pp. 40-43.

201 Bonnie Sherk, quoted in: Ruth Iskin, "The Fine Art of Social Change," *Social Works*, LAICA, Los Angeles, 1979, p. 10.

202 On the maze, see Ronald Onorato, "The Modern Maze," *Art International*, April/May 1976, pp. 21-25; Hermann Kern, "Labyrinths: Tradition and Contemporary Works," *Artforum*, May 1981, pp. 60-68; On gardens and pathways see Lucy R. Lippard, "Gardens:Some Metaphors for a Public Art," *Art in America*, November 1981, pp. 136-150.

203 During the latter half of the sixties there was also in Los Angeles the tendency to move away from the solid art object. A small group of artists here had a special interest in investigating the characteristics of light and light phenomena: Larry Bell, Robert Irwin, Eric Orr, Maria Nordman, Douglas Wheeler, and James Turrell. Before turning to sound installations, Michael Brewster's works were pieces of flickering light bulbs, which he placed in different configurations in the Mojave Desert (California).

204 The symposium took place in the former Malinda Wyatt Gallery on Market Street, Venice, California. See Jane Livingston, *A Report on the Art and Technology Program*, pp. 140-142. See also a recent publication on the subject by Jan Butterfield, *Art of Light and Spaces*, Abbeville Press, New York 1993.

205 James Turrell, *Light Projections and Light Spaces*, Stedelijk Museum, Amsterdam 1976.

206 Graig Adcock, "Light, Space, Time: The Visual Parameters of the Roden Crater," *Occluded Front: James Turrell*, The Museum of Contemporary Art, The Lapis Press, Los Angeles 1985, pp. 101-135. The catalogue contains a full biography and bibliography until 1985. Also: Graig Adcock, *James Turrell: The Art of Light and Space*, University of California Press, Berkeley/Los Angeles, 1992.

207 Janet Saad-Cook, "Touching the Sky: Artworks using Natural Phenomena, Earth, Sky and Connections to Astronomy," *Leonardo*, Vol. 21, No. 2, 1988, p. 129; Graig Adcock, "Anticipating 19,084: James Turrell's Roden Crater Project," *Arts Magazine*, May 1984, pp. 76-85.

208 Krupp studied the astronomy of ancient peoples. See E.C. Krupp, ed., *In Search of Ancient Astronomics*, Doubleday, Garden City, NY 1977.

209 James Turrell: "...Roden Crater has knowledge in it, and it does something with that knowledge. Environmental events occur; a space lights up. Something happens in there, for a moment, or for a time. It is an eye, something that is itself perceiving. It is a piece that does not end. It is changed by the action of the sun, the moon, the cloud cover, by the day and the season that you're there...and it keeps changing. When you're there, it has visions, qualities, and a universe of possibilities." In: Graig Adcock, "Light, Space, Time: The Visual Parameters of the Roden Crater," *Occluded Front: James Turrell*, p. 130.

210 Janet Saad-Cook, "Touching the Sky: Artworks using Natural Phenomena, Earth, Sky and Connections to Astronomy," p. 127.

211 Nancy Holt, "Stone Enclosure: Rock Rings, 1977-78," *Arts Magazine*, June 1979, pp. 152, 153. Other publications by the artist: "Hydra's Head," *Arts Magazine*, January 1975, pp. 57-59; "Pine Barrens," *Avalanche*, Summer 1975, p. 6; "Sun Tunnels," *Artforum*, April 1977, pp. 32-37; Ransacked, Printed matter, Inc., New York 1980; "Pipeline," *Alaskan Impressions*, Anchorage, AL 1986 (includes selected bibliography).

212 Janet Saad-Cook, "Touching the Sky: Artworks using Natural Phenomena, Earth, Sky and Connections to Astronomy," p. 127.

213 Nancy Holt, quoted in *Art in America*, February 1991, p. 79.

214 As these projects grow in size, so are the costs. The $600,000 range is by no means uncommon for the design of a park landscape now. In the past the costs of her projects ranged from a mere $200 for *Polar Circle* (Miami University, Oxford, Ohio, 1979) - which was built from found poles and army surplus chain, while students assisted with the construction, - to $50,000 - $100,000. It shows how different the production of this type of work is from that of the painter in his studio.

[215] Barbara C. Matilsky, *Fragile Ecologies, Contemporary Artists' Interpretations and Solutions*, Rizzoli International Publications, Inc., in association with The Queens Museum of Art, New York 1992, pp. 90-91.

[216] About Irwin's first major show at the Los Angeles County Museum of Art in 1966, Philip Leider wrote: "In Robert Irwin's most recent paintings one is confronted by what appears to be an immaculate white picture plane, about seven feet square, and nothing more. Some time must pass - a minute or two, or three - before the viewer becomes fully aware of an indistinct, irregularly shaped mass which seems to have emerged out of the white plane...roughly centered. The coloration is so subdued that there is no possibility of defining what one sees in terms of it, but rather in terms of what it suggests: a quality of energy, an energy, one feels, which will tend, ultimately to dissolve itself uniformly on the picture plane in a kind of entropic dissipation." In: Philip Leider, *Robert Irwin/Kenneth Price*, Los Angeles County Museum of Art. Los Angeles 1966, n.p.

[217] See John Coplans, *Robert Irwin*, Pasadena Art Museum, Pasadena 1968, n.p.; Jan Butterfield, "Part I. The State of the Real: Robert Irwin discusses the Activities of an Extended Concsiousness," *Arts Magazine*, June 1972, pp. 47-49.

[218] For a chronology and bibliography, see *Robert Irwin*, The Whitney Museum of Art, New York 1977; *Robert Irwin*, The Museum of Contemporary Art, Los Angeles, Rizzoli International Publications, New York, 1993.

[219] *The last Time I saw Ferus: 1957-1966*, Newport Harbor Art Museum, Newport Beach, CA 1976. Essay by Betty Turnbull.

[220] Roberta Smith, "Robert Irwin: The Subject is Sight," *Art in America*, March/April 1976, p. 72.

[221] Ira Licht, "Robert Irwin," *Robert Irwin*, Museum of Contemporary Art, Chicago 1975, n.p.

[222] The recent Robert Irwin retrospective at The Museum of Contemporary Art in Los Angeles was not without contradictions, Irwin admittted, and in fact totally inconsistent with his current theories and works. See *Robert Irwin*, The Museum of Contemporary Art, Los Angeles, Rizzoli International Publications, New York 1993, p. 181.

[223] 1. Site dominant. This work embodies the classical tenets of permanence, transcendent and historical content, meaning and purpose; the art-object either rises out of, or is the occasion for, its "ordinary" circumstances - monuments, historical figures, murals, etc. 2. Site adjusted. Such work compensates for the modern development of the levels of meaning-content having been reduced to terrestrial dimensions (even abstraction). Here consideration is given to adjustments of scale, appropriateness, placement, etc. But the 'work of art' is still either made or conceived in the studio... 3. Site specific. Here the 'sculpture' is conceived with the site in mind; the site sets the parameters and is, in part, the reason for the sculpture. This process takes the initial step towards sculpture's being integrated into its surroundings. But our process of recognition and understanding of 'the work of art' is still keyed to the oeuvre of the artist. 4. Site conditioned/determined. Here the sculptural response draws all of its cues (reasons for being) from its surroundings. This requires the process to begin with an intimate hands-on reading of the site. This means sitting, watching, and walking through the site, the surrounding areas... In: Robert Irwin, *Being and Circumstance*, The Lapis Press, in conjunction with The Pace Gallery, New York and the San Fransisco Museum of Modern Art, San Francisco 1985, pp. 26-28.

[224] Robert Irwin, *Being and Circumstance*, pp. 26-28.

[225] Robert Irwin, *Robert Irwin*, The Museum of Contemporary Art, Los Angeles, Rizzoli International Publications, New York 1993, p. 20.

[226] Lawrence Weschler, *Seeing is Forgetting the Name of the Thing One Sees: A Life of Contemporary Artist Robert Irwin*, University of California Press, Berkeley/ Los Angeles 1982, p. 130-1. Biography on the artist's life and work.

[227] Robert Campbell & Jeffrey Cruikshank, "Art in Architecture," *Designing the Wiesner Building*, Massachusetts Institute of Technology, Cambridge, MA (n.d.), p. 86.

[228] Lawrence Weschler, "Playing It As It Lays & Keeping It In Play: A Visit with Robert Irwin," *Robert Irwin*, 1993, pp. 158-168, p. 159.

[229] Ibid., *Robert Irwin*, p. 151.

[230] Ibid., *Robert Irwin*, "The Hidden Structures of Art," p. 35.
[231] Lawrence Weschler, from the UCLA Library Oral History Program, edited interview by Frederick Wight with Robert Irwin in 1976 in "Robert Irwin's Whitney Project - Retrospects and Prospects," *Laica Journal*, No. 15, July/August 1977, p. 19.
[232] Lawrence Alloway, *An Autobiography of Alan Sonfist*, Herbert F. Johnson Museum of Art, Ithaca, NY 1975, n.p. See also his statement in: Alan Sonfist, *Army Ants: Patterns and Structures*, Automation House, New York 1972 (n.p.).
[233] *Elements: Earth, Air, Fire, Water*, Boston Museum of Fine Arts, Boston 1971.
[234] Jonathan Benthall, "Haacke, Sonfist and Nature," *Studio International*, March 1971, pp. 95-96.
[235] Alan Sonfist, *Elements: Earth, Air, Fire, Water*, p. 54.
[236] Grace Glueck, "Nature's Boy," *The New York Times*, November 15, 1970.
[237] Michael Auping, "Interview with Alan Sonfist," *Common Ground: Five Artists in the Florida Landscape*, The John and Mable Ringling Museum of Art, Sarasota, FL 1982, p. 134.
[238] Alan Sonfist, statement at the presentation of *Landscapes: Crystals-Microorganisms, Multiple Interaction Team*, CAVS/MIT Press, Cambridge 1972, n.p.
[239] Robert Joseph Horvitz, "Nature as Artifact: Alan Sonfist," *Artforum*, November 1973, p. 32.
[240] Ibid., p. 33.
[241] Statement published with the exhibition of *Animal, Vegetable, and Mineral Substances Landscapes*, Paley and Lowe Gallery, New York 1972.
[242] Alan Sonfist, *Natural Phenomena as Public Monuments*, Neuberger Museum, State University of New York, Purchase, NY 1978, p. 2.
[243] Michael Auping, "Interview with Alan Sonfist," *Common Ground: Five Artists in the Florida Landscape*, p. 134.
[244] Michael Heizer, *Effigi Tumuli: The Reemergence of Ancient Mound Building*, Harry N. Abrams, New York 1990.
[245] Alan Sonfist, *Multiple Interaction Team*, n.p.
[246] Alan Sonfist, *Natural Phenomena as Public Monuments*, p. 5.
[247] Jeffrey Deitsch, "The New Economics of Environmental Art," *Art in the Land*, edited by Alan Sonfist, E.P. Dutton, Inc., New York 1983, p. 90.
[248] For bibliographical data, until 1982, see Michael Auping, *Common Ground: Five Artists in the Florida Landscape*, pp. 142-145.
[249] Ken Johnson, "Alan Sonfist at Ledisflam," *Art in America*, January 1992, p. 109.
[250] Peter Birmingham, *Alan Sonfist/Trees*, Smithsonian Institution, Washington D.C. 1978, n.p.
[251] Jack Burnham wrote of Sonfist's work very aptly and succinctly: "He sees 'sculpture' in the ecological exchanges that occur every day and believes that man-nature stability will come only when we have become acutely sensitive to the natural changes around us." In: Jack Burnham, "Art and Technology," *Britannica Yearbook of Science and the Future*, University of Chicago Press, Chicago, 1973, p. 351.
[252] Alan Sonfist, *Natural Phenomena as Public Monuments*, p. 8.
[253] The *Report* mentions the background of this interest at length. See *A Report on the Art and Technology Program*, p. 118.
[254] Gail Scott, "Newton Harrison," ibid., p. 122.
[255] Graig Adcock, "Conversational Drift: Helen Mayer Harrison and Newton Harrison," *Art Journal*, Summer 1992, pp. 36-36.
[256] Michael Auping, "Interview with Helen and Newton Harrison," *Common Ground: Five Artists in the Florida Landscape*, p. 42 .
[257] Jonathan Benthall, "Newton Harrison: big fish in small pool," *Studio International*, December 1971, p. 230.
[258] Kay Larson, "Watering Wholes," *Village Voice*, April 14, 1980, p. 78.
[259] An eco-system is defined as: "A system...of interacting subsystems which persists through time due to the interaction of its components. The system possesses a definable organization, temporal continuity and functional properties which can be viewed as distinctive to a system rather that its components...a system is specified as a complex unit in which the components interact to

preserve the sytem and restore it following non-destructive disturbances." In: W.H. van Dobben & R.H. Lowe-McConnell, Unifying Concepts in Ecology: Report of the Plenary Sessions of the First International Congress of Ecology, The Hague, September 8-14, 1974, pp. 27-28.

[260] *The Lagoon Cycle*, Herbert F. Johnson Museum, Cornell University, Ithaca and New York 1985.

[261] Kay Larson, "Watering Wholes," p. 78.

[262] Helen Mayer Harrison and Newton Harrison, "Shifting Positions toward the Earth and Environmental Awareness," *Leonardo*, Vol. 26, No. 5, 1993, pp. 374.

[263] Michael Auping, "Helen and Newton Harrison," *Common Ground: Five Artists in the Florida Landscape*, p. 45.

[264] Grace Glueck, "Art People: 'The Earth is Their Palette'," *The New York Times*, 4 April, 1980.

[265] Newton Harrison, "Sea Grant second narrative and two precedent works," *Studio International*, May 1974, p. 236.

[266] Later on Newton Harrison wrote in defense of *Portable Fish Farm* that "it was first conceived to sit outdoors. It used earth and water pastures with minimal support systems. Lack of time and funding problems reduced it exclusively for the indoor gallery." In: Newton Harrison, "Sea Grant second narrative and two precedent works," p. 237.

[267] Critic Kristine Stiles heavily criticized the artists upon their presentation of the *Meditations on the Condition of the Sacramento River, the Delta, and the Bays of San Francisco*, stating that they "neglected to establish a network of communication feedback and tabulation which would measure and thus ensure the work's effectiveness in the San Francisco community.." In: Kristine Stiles, "Helen and Newton Harrison: Questions," *Arts Magazine*, February 1978, pp. 131-133.

[268] *The Serpentine Lattice, A Conceptual Design and Installation by Newton Harrison and Helen Mayer Harrison.* The Douglas F.Cooley Memorial Art Gallery, Reed College, Portland, Oregon and Ron Feldman Fine Arts, New York 1993.

[269] Graig Adcock, "Conversational Drift: Helen Mayer Harrison and Newton Harrison," pp. 40-41.

[270] From the brochure *Guadelupe Meander, A Refugia for San Jose*, San Jose State University, San Jose, CA 1983.

[271] Helen Mayer Harrison and Newton Harrison, "Shifting Positions Toward the Earth: Art and Environmental Consciousness," p. 376.

[272] *Atempause für den Save Fluß, Die Summe seiner Geschichte, Beginn einer neuen Geschichte*, Neuer Berliner Kunstverein in Zusammenarbeit mit dem Berliner Künstlerprogramm des DAAD, Berlin, Germany and Galerie Ljubljana, Slovenia, 1989.

[273] *Future Garden: Teil I. Die gefährdeten Wiesen Europas*. Helen Mayer Harrison, Newton Harrison, und Harrison Studio, Kunst-und Ausstellungshalle der Bundesrepublik Deutschland, Bonn 1996.

[274] Newton Harrison and Helen Mayer Harrison, *Tibet is the High Ground*, (Del Mar), 1990, pp 1, 6. Selected finished projects : *Baltimore Promenade*, Baltimore 1981 (large sections); *Devil's Gate Transformation: A Refuge for Pasadena*, Pasadena, California,1987 (under construction); *A Wetland Walk for Boulder Creek*, Boulder, Colorado, 1987; *Disappearing Path*, Newport, California, 1988; *San Diego Landfill*, together with Martinez, Cutri and McArdle, San Diego, California,1992; *California Wash*, Santa Monica, California, 1996.

[275] Graig Adcock, "Converational Drift: Helen Mayer Harrison and Newton Harrison," p. 43.

[276] Susan Platt, "Helen Mayer Harrison and Newton Harrison: An Urban Discourse," *Artweek*, May 21, 1983, p. 1, backpage.

[277] A listing of artists regularly working in this field would be rather extensive. Among the better-known currently creating permanent public works, and not discussed in this book, are: Robert Basler, Carl Cheng, Michael Davis, Richard Fleischner, Ian Hamilton-Finley, Lloyd Hamrol, Hera, Mary Miss, Max Neuhaus, Claes Oldenburg/Coosje van Bruggen, Beverly Pepper, Elyn Zimmerman.

[278] Robert Irwin, *Being and Circumstance*, p. 49.

[279] *Lloyd Hamrol: Works, Projects, Proposals*, Interview with Lloyd Hamrol by Julia Brown Turrell, Municipal Art Gallery, Los Angeles 1986, pp. 42-49.

[280] A few small and early reclamation projects should be mentioned. They were part of the *Program in Visual Arts* at Artpark, Lewiston, New York, which has invited artists-in-residence annually since 1974 during the summer months. Artists Charles Simonds (1975), Alan Sonfist (1975) and Helen and Newton Harrison (1977) worked here on projects reclaiming parts of a chemical dump site that belonged to the Artpark area. New York-based artist Charles Simonds created a *Growth House* (1975) on the same chemical dump site. The *Growth House* was a circular structure built from sacks of earth containing a large variety of seeds, which were to sprout at various times of the year, forming a complete life cycle of food-producing plants - grains, vegetables, fruits, herbs - throughout the year. And Alan Sonfist had used part of the spot for his *Pool of Virgin Earth* in the same year. Sonfist made a circular pool covered with virgin soil to catch seeds from neighboring areas and thus start the rebirth of a new forest. See *Artpark: The Program in the Visual Arts*, Lewiston, NY 1975, 1976, 1977.

[281] *Earthworks: Land Reclamation as Sculpture,* Seattle Art Museum, Seattle, WA 1979, p. 5.

[282] See *Earthworks: Land Reclamation As Sculpture - Technical Report,* King County Arts Commission, Seattle, WA 1981, for a complete listing of funds and budgets.

[283] Already in 1955, Herbert Bayer had created an *Earth Mound* and *Marble Garden* for the Aspen Institute of Humanistic Studies in Colorado, where he held a position. Although Robert Smithson is considered the contemporary theoretical father of this direction, Herbert Bayer and Isamu Noguchi, who have been proposing environmental works since 1933, are the actual forerunners of the current recultivation - landscaping trend. Apart from his sculptural works, Isamu Noguchi has designed numerous urban plazas with multiple functions.

[284] Patricia Johanson, *Fragile Ecologies,* p. 82.

[285] Agnes Denes, "Environmental Artwork, Visual Philosophy and Global Perspective," *Leonardo,* Vol. 26, No. 5, 1993, p. 392.

[286] She has done environmentally oriented work since the seventies, at first mainly temporary, and has written about her concerns and visions in periodicals and books. See Agnes Denes, *The Beginning and the End of Time and Thereafter,* Visual Studies Workshop, Rochester, NY 1989.

[287] Los Angeles Community Redevelopment Agency, *Public Art Policy,* Los Angeles 1993.

[288] Marga Bijvoet, "Underground Spectacle in Los Angeles," *Archis,* No. 3, 1994, pp. 72-80.

[289] Douglas C. McGill, "Architect and Artists Collaborate on Battery Park City Plaza," *New York Times,* January 31, 1985, III, p. 17.

[290] Siah Armajani, statement in: Shaping the New Sculpture of the Street, a symposium, *New York Times,* September 22, 1985. Also: Ken Johnson, "Poetry & Public Service," *Art in America,* March 1990, pp. 161-163, 219.

[291] Howard Wise Gallery in New York (later on Electronic Arts Intermix) organized *TV as a Creative Medium* in 1969. In 1970, the Rose Art Museum, at Brandeis University (Waltham, MA) organized the first video exhibition *Vision and Television.* The Everson Museum of Art in Syracuse established a major center for video art for years to come, followed by the Museum of Modern Art in New York, and the Long Beach Museum of Art in Long Beach.

[292] David Bienstock, program notes for *A Special Videotape Show,* Whitney Museum of American Art, New York 1971, n.p.

[293] René Berger, "Video and the Restructuring of Myth," *The New Television: A Public / Private Eye,* ed. Douglas Davis/Allison Simmons, MIT Press, Cambridge, MA 1978, pp. 206-221.

[294] Wulf Herzogenrath and Edith Decker, eds. *Video Skulptur: retrospektiv und aktuell,* DuMont-Dokumente, Cologne 1989.

[295] Nam June Paik, "New Projects 1972/73," *Nam June Paik: Videa 'n' Videology (1959-1973),* Everson Museum of Art, Syracuse, NY 1974, p. 75. The pagination made afterwards begins with the foreword. According to this pagination, quote on p. 1. This catalogue is still the most extensive resource for Paik's letters, documents, writings until 1973. It is recently supplemented by: Nam June Paik, *Niederschriften eines Kulturnomaden,* ed. Edith Decker, DuMont, Cologne 1992.

[296] James Harithas, "Introduction," *Nam June Paik: Videa 'n' Videology (1959-1973),* 1974

[297] John Cage, *Silence*, Weslyan University Press, Middletown, CT 1961, p. 40.

[298] *Frankfurter Allgemeine Zeitung*, Feuilleton, No. 280, 2.12.1961.

[299] Nam June Paik, "Norbert Wiener and Marshall McLuhan," *Institute of Contemporary Arts Bulletin*, London 1972, No. 3, p. 3 ff. Reprinted in *Videa 'n' Videology*, 1974, pp. 7-9. In McLuhan's thesis all techniques and technologies were "extensions of man" and never neutral in their effects on the existing environment. *The Gutenberg Galaxy* (1961) and *Understanding Media: the extensions of man* (1964) were widely read. He was both hailed as the new spokesman of the electronic age and vehemently criticized., particularly by the scientific world, as his methods were hardly empirical. See Jonathan Miller, *McLuhan*, London 1971. Recently, Marshall McLuhan's ideas are going through some revaluation. In 1987 an edition of letters was published, and in 1988 his son Eric McLuhan published *Laws of the Media*, posthumously.

[300] Jud Yalkut, "Art and Technology of Nam June Paik," *Arts Magazine*, April 1968, pp. 50-51. See also Nam June Paik, "Simulation of Human Eyes by 4-Channel Stereo Video Taping," *Videa 'n' Videology*, 1974, p. 62.

[301] Nam June Paik, "Random Access Information," *Artforum*, December 1980, pp. 46-49.

[302] Jack Burnham, *Beyond Modern Sculpture*, pp. 351-352 .

[303] *McLuhan Caged* was exhibited in Pontus Hultén's *The Machine as Seen at the End of the Mechanical Age*, Museum of Modern Art, New York 1968, together with a piece titled *VRondo Electrique*, a manipulated television set with dancing wave patterns.

[304] Extensive literature on the early aspects of Paik's oeuvre can be found in Wulf Herzogenrath, *Nam June Paik, Fluxus Video* (Munich. 1983); *Nam June Paik, Werke 1946-1976, Musik-Fluxus-Video* (Kölnischer Kunstverein, Cologne 1976). Edith Decker, *Paik Video*, gives a survey of installations and videotapes (DuMont, Cologne 1988). The book contains full biographical and bibliographical information until 1988.

[305] See for other "visionary" statements on the future of electronic communication systems, in the television era a letter to *Radical Software*, the "Binghamton Letter," January 8, 1972, reprinted in *Videa 'n' Videology*, 1974, pp. 68-69; Nam June Paik, "Medienplanung für das nachindustrielle Zeitalter (1974) Bis zum 21.Jahrhundert sind es nur noch 26 Jahre," *Nam June Paik, Werke 1946-1976, Musik-Fluxus-Video*, 1976, p. 159.

[306] Nam June Paik, "New Projects 1972/73," *Videa 'n' Videology*, 1974, p. 76.

[307] Nam June Paik, "Global Groove and Video Common Market," *The WNET-TV Lab News*, Issue #2, 1973 (written February 1970). Reprinted in *Videa 'n' Videology*, 1974, pp. 60-61.

[308] Nam June Paik, "Expanded Education for the Paperless Society," *Videa 'n' Videology*, 1974, pp. 31-37.

[309] Nam June Paik, *Videa 'n' Videology*, 1974, p. 11.

[310] Published in German in 1976 in: *Nam June Paik, Werke 1946-1976. Musik-Fluxus-Video*, ed. Wulf Herzogenrath, Kölnischer Kunstverein, Cologne 1976, pp. 165-166.

[311] Nam June Paik, in: Catalogue Biennale di Venezia, Venice 1993, pp. 176-177.

[312] Karl Mannheim, *Ideologie und Utopie*, Frankfurt 1969, pp. 171 ff.

[313] Leaflet: *Café Au Go Go*, 152 Bleecker Street - October 4 & 11 1965 - a preview of Paik's November show at the Bonino Gallery. Reprinted in *Videa 'n' Videology*, p. 11.

[314] Nam June Paik, "New Projects 1972/73," *Videa 'n' Videology*, p. 76.

[315] "...e.g. Mike Noll (Bell Labs) last month c.cmpleted a 3-D figuration using the light pen, cathode ray tube, and medium sized computer. It can be useful for experimental art classes, may be for sculpture class, and drawing courses for engineering students and the teaching of solid geometry... I personally feel that high level applications like (Michael) Noll's and (Max) Mathew's programming would be more valuable, since it means the augmentation of new knowledge, and the creation of new art, whereas low level applications such as sight-singing, or ear-training is just a money saving in teaching yesterday's music..." In: Nam June Paik, "Expanded Education for the Paperless Society," *Videa 'n' Videology*, 1974, pp. 31-37. Paik knew Billy Klüver and his E.A.T. organization, and thus the way to Bell Laboratories was paved.

[316] Jud Yalkut, "Art and Technology of Nam June Paik," p. 51.

[317] Nam June Paik, "De-Composition in the Media Art," *Nam June Paik: eine DATA base*, ed. Klaus Bußmann / Florian Matzner, Edition Cantz, 1993, pp. 18-19, 21.

[318] Leaflet *Nam June Paik*, Whitney Museum of American Art, New York 1982. Also: John G. Hanhardt, *Nam June Paik*, Whitney Museum of American Art, New York 1982.

[319] "My obsession with TV for the past ten years has been, if I look back and think clearly, a steady progrssion towards more differentiated participation by viewers," he stated in the catalogue of the Sonsbeek exhibition. Nam June Paik, *Sonsbeek 71*, Arnhem, NL 1971, p. 84.

[320] Nam June Paik, "About the Exposition of Electronic Music," *De-collage*, 1962, No. 3, n.p.

[321] There is a description by critic Tomas Schmitt in: "exposition of music," *Nam June Paik, Werke 1946-1976 Musik-Fluxus-Video*, p. 68.

[322] Douglas Davis, "Nam June Paik: The Cathode Ray Canvas," *Art and the Future*, pp. 146-152.

[323] "Participation TV II included three or four color TV sets which show multi-color echoes, or fog, or clouds which are electronically produced. Sometimes you can see yourself floating in air, dissolving in deep water," wrote Paik for the accompanying brochure. Howard Wise Gallery, *TV as a Creative Medium*, New York May 17 - June 14, 1969, n.p.

[324] David Antin analyzed the formal and sequential structure of the television image, its fixed and preset timing and pacing as applied by the American networks in detail in an article entitled "Television: Video's Frightful Parent," *Artforum*, December 1975, pp. 36-45.

[325] David A. Ross, "Nam June Paik's Videotapes," *Nam June Paik*, Whitney Museum of American Art, New York 1982, pp. 101-110.

[326] Among the public television stations KCET-TV in San Francisco was the first on to open its doors to experiments by artists (1967) offering residencies for artists, followed by WGBH-TV in Boston (1968), where Stan VanderBeek was the first artist-in-residence. Nam June Paik became the second one.

[327] Douglas Davis, "Nam June Paik: The Cathode Ray Canvas," *Art and the Future*, pp. 146-154.

[328] Edith Decker, *Paik Video*, pp. 108-111.

[329] Kenneth E. Silver, "Nam June Paik: video's body," *Art in America*, November 1993, p. 105.

[330] The whole program was shown later on in an exhibition organized by DAAD, in Berlin and was accompanied by a extensive catalogue, containing documentary information, essays, comments and critiques about the program, titled: *Art for 25 Million People*, DAAD, Berlin 1984.

[331] See for a description of Nam June Paik's videotapes, Edith Decker, *Paik Video*, 1988, David A. Ross, "Nam June Paik's Videotapes," *Nam June Paik*, 1982, pp. 101-110; Bruce Kurtz, "Paikvision," *Artforum*, February 1982, pp. 52-55.

[332] J. Hoberman, "The Reflecting Pool: Bill Viola and the Visionary Company," *Bill Viola*, Museum of Modern Art, New York 1987, p. 64.

[333] Calvin Tomkins, "Video Visionary," *The New Yorker*, 5 May, 1975, p. 45.

[334] Raymond Bellour, "An Interview with Bill Viola," *October*, 34, Fall 1985, p. 101.

[335] Bill Viola, statement, *Bill Viola*, ARC Musée d'Art Moderne de la Ville de Paris, Paris 1983, p. 4. Also: Bill Viola, "Video As Art," *Video Systems*, July 1982, pp. 26-35.

[336] Michael Nash, "Bill Viola's Re-Visions of Mortality," *High Performance*, 37, 1987, p. 63.

[337] Bill Viola, "Excerpt from The Sound of One Line Scanning," *1986 National Video Festival*, American Film Institute, Los Angeles 1986, p. 40. Bill Viola's writings have been collected in: *Reasons for Knocking at an Empty House: Writings 1973-1994*, ed. Robert Violette in collaboration with the author. Cambridge, MA: The MIT Press; London: Thames & Hudson, in association with Anthony D'Offay Gallery, 1995.

[338] Bill Viola, statement in *Bill Viola*, Paris 1983, p. 31.

[339] Marita Sturken, "The Videotapes of Bill Viola," *Afterimage*, Summer 1982, p. 28.

[340] Bill Viola, *Bill Viola*, New York 1987, p. 37.

[341] Taped interview with the artist, March 25, 1987.

[342] Bill Viola, "Excerpt from The Sound of One Line Scanning," p. 43.

[343] From "The Vasulkas," *Cantrills Filmnotes*, No. 13, April 1973.

[344] Bill Viola, "Excerpt from The Sound of One Line Scanning," p. 42.

[345] Syring, Marie Luise, ed., *Bill Viola: Unseen Images*, Kunsthalle, Düsseldorf 1992.

[346] Deirdre Boyle, Interview of Bill Viola, *Bill Viola*, Paris 1983, p.14.

[347] Bill Viola, in *Bill Viola. Buried Secrets*, ed. Marilyn A. Zeitlin, Arizona State University Art Museum, Tempe, AZ and Kestner Gesellschaft, Hannover, Germany, p. 64.

[348] Bill Viola, statement in *Bill Viola*, New York 1987, p. 55.

[349] Sogyal Rinpoche, *The Tibetan Book of Living and Dying*, Harper San Francisco, p. 37.

[350] Michael Nash, "Bill Viola's Re-Visions of Mortality," p. 63.

[351] The Rig Vedas were written in Sanskrit, the language brought to India during the first half of the second milllennium before Christ, to a region now called the Punjab. The Rig Vedas are the result of a systematic collection of versions originally spread over the subcontinent. Central to the Rig Vedas is the uncertainty of knowledge of nature's laws and phenomena.

[352] Kathy Rae Huffman, cover text, accompanying the 1/2" tape.

[353] Taped interview with the artist, March 25, 1987.

[354] Quoted from "Putting the Whole back Together," Bill Viola in conversation with Otto Neumaier and Alexander Pühringer. In: *Bill Viola*, ed. Alexander Pühringer, Salzburg: Verlag Ritter Klagenfurt/Salzburger Kunstverein 1994, p. 138.

[355] Quotes from: *Bill Viola: Buried Secrets*, pp. 27-31.

[356] Marilyn A. Zeitlin, "Continuities in the Work of Bill Viola," *Bill Viola. Buried Secrets*, p. 58.

[357] Paul Ryan, *Video Mind, Earth Mind: Art, Communication and Ecology*, Peter Lang Publishing, New York 1992.

[358] Paul Ryan, "A Genealogy of Video," *Leonardo*, Vol. 21, No. 1, 1988, p. 39.

[359] John Giancola is currently preparing an ethnography of this early period of video work. Deirdre Boyle's *Video Classics* includes a number of the documentary/art videotapes made in the early seventies, but does not provide a survey. See Deirdre Boyle, *Video Classics*, Oryx Press, Phoenix, AZ 1986.

[360] Willoughby Sharp, and Paul Ryan, "From Teenage Monk to Video Bareback Rider," A Willoughby Sharp Interview, *Video 81*, 2 (1) pp. 14-17.

[361] Paul Ryan, "A Genealogy of Video," p. 40.

[362] Jud Yalkut, Interview with Frank Gillette and Ira Schneider, *East Village Other*, August 6, 1969. Also: Gene Youngblood, *Expanded Cinema*, E.P. Dutton & Co., Inc., New York 1970, pp. 341-343.

[363] James Harithas, "Blueprint for a recreative reorientation," *Video: Process and Meta-process*, Everson Museum of Art, Syracuse, NY 1973. Essay by Frank Gillette. Reprinted in *Radical Software*, Vol. 2, No. 5, 1973, pp. 44-45.

[364] Raindance Archives, 51 Fifth Avenue, New York City. Electronic Arts Intermix, New York, has some Raindance tapes in distribution.

[365] Global Village, founded by John Reilley and Julie Gustaffson in 1969, still produces independent video documentaries, schedules workshops and seminars, has regular screenings, and functions as a video study center. Video-freex was begun by David Cort. The Alternative Media Center at New York University was started by George Stoney and Red Burns, and holds an important archive of documentary video (many of it videotaped by artists). People's Video Theatre was initiated by the painters Ken Marsh and Howard Gutstadt.

[366] Paul Ryan, "Cable Television: The Raw and the Overcooked," *Radical Software*, No. 1, New York, 1970, p. 12.

[367] Michael Shamberg, "Guerilla Television," *Radical Softrware*, Vol. 2, No. 1, September 1972, editorial page.

[368] McLuhan had just published his book *War and Peace in the Global Village*, New York 1968.

[369] Paul Ryan, "Cybernetic Guerilla Warfare," *Radical Software*, No. 3, Spring 1971, pp. 1-2.

[370] Paul Ryan, "The Earthscore Notational System for Orchestrating Perceptual Consensus about the Natural World," *Leonardo*, Vol. 24, No. 4, 1991, p. 457.

[371] Paul Ryan, *Video Mind, Earth Mind*, p. 1.

[372] Charles S. Peirce, *The New Elements of Mathematics*, Collected Papers, Vol. IV, ed. C. Hawthorne / P. Weiss, Cambridge, MA, 1982, p. 339.

[373] Gregory Bateson, *Steps to an Ecology of Mind*, p. 441.

[374] Ibid., p. 483.
[375] Ibid., p. 487.
[376] Warren McCulloch, *Embodiment of Mind*, MIT Press, Cambridge, MA 1965.
[377] Paul Ryan, "Attempting a Calculus of Intention," *Radical Software*, Vol. 1, No. 3, 1971; Paul Ryan, "Toward an Information Economy," *Radical Software*, Vol. 1, No. 3, 1971.
[378] Paul Ryan, "The Millennium, Montage, Remote Sensing and the Earthscore Method," in *Codes and Customs: Millennial Perspectives*, ed. Roberta Kevelson, Peter Lang Publishing, New York 1994, p. 261.
[379] Paul Ryan, "The Earthscore Notational System for Orchestrating Perceptual Consensus about the Natural World," p. 464. Ryan's "Earthscore Notational System for Orchestrating Perceptual Consensus about the Natural World," addresses different layers of his system, theoretically and applied to video, as well as difficulties involved in creating a consensus concerning local ecologies in relation to language and perception. It also states Paul Ryan's (philosophical) theoretical position.
[380] Ibid., pp. 457-465.
[381] Paul Ryan, "The Millennium, Montage, Remote Sensing and the Earthscore Method," p. 245.
[382] In the early seventies the idea of utopian community living within ecological limits had become quite fashionable. And Ernest Callenbach's *Ecotopia*, published in 1975 was still a widely read book. However, Ryan says that he never cared much for this direction, nor owed any ideas to it.
[383] Paul Ryan, *Video Mind, Earth Mind*, Part Three: Bioregions, 1977-1980, pp. 158-59.
[384] James Lovelock, *Gaia: A New Look at Life on Earth*, Oxford University Press, London and New York 1995 (Revised version).
[385] There were four quarterly issues published, totalling a distribution of 15,000 copies. *Talking Wood* was set up as a non-profit organization, and financed by the Labor Department's Comprehensive Employment and Training Act (CETA).
[386] Paul Ryan, *Video Mind, Earth Mind*, Part Three: Bioregions, 1977-1980, p. 159.
[387] Paul Ryan, *Video Mind, Earth Mind*, Part Four: Ecochannel, 1981-1985, p. 202. The Ecochannel Design was presented in lectures and videotapes at the Museum of Modern Art, New York, in 1985, the first International EcoCity Conference, and and the World Congress of Local Governments for a Sustainable Future at the United Nations. It was published as "The Ecochannel Design," *IS Journal*, 2 (2), 1987, pp. 46-64.
[388] Ibid., p. 240.
[389] Ibid., p. 290.
[390] Ibid., p. 293. The monitoring quality of television is extensively discussed in a forthcoming article "Neoluddites: Usurpers of Our Electronic Future."
[391] Director/author is Jean Gardner; another member with whom Ryan works is artist/composer David Dunn. The Earth Environmental Group is a non-profit arts and education collaborative, dedicated to increasing public awareness of natural and architectural environments. Over the past twelve years, the collaborative carried out extensive field investigation and research into New York City's bioregions. Studies were conducted on vegetation, geology, wildlife, natural history, and human habitat. The findings of the Earth Environmental Group have been presented to the public in pamphlets, posters, filmstrips, radio programs, video, multimedia shows, and workshops.
[392] Paul Ryan, "The Millennium, Montage, Remote Sensing and the Earthscore Method," p. 260.
[393] Letter from the artist, February 12, 1996.
[394] *Vasulka. Steina: machine vision/Woody: descriptions*, ed. Linda Carthcart, Albright-Knox Art Gallery, Buffalo 1978, pp. 23, 31. See also: Wolfgang Preikschat, The Vasulkas - Vom Video-feedback zur hybriden Interaktion," *Kunstforum International*, March 1992, pp. 184-193.
[395] Woody Vasulka in the videotape *Artifacts*, 1980.
[396] NYCA files. The Vasulkas were funded as members of a foundation, "Perception" in 1971 and 1972 through Howard Wise's Intermix (the future Electronic Arts Intermix).
[397] The Kitchen opened June 15, 1971. Vasulkas: We said, "Let's use it electronically, as a place that does electronic music and electronic sounds... We, in fact, enjoyed certain things that were

forbidden to the true radicals in the sense of purity of thinking of Buckminster Fuller or McLuhan. We would be very much involved in the phenomenon of time. And we could incorporate all those things: we took a certain interest in that particular aspect." In: *Vasulka. Steina: machine vision/Woody: descriptions*, p. 32.

398 David L. Shirey, "Video Art Turns to Abstract Imagery," *The New York Times*, July 4, 1972, p. 6.

399 Steina Vasulka, in Lucinda Furlong, "Notes Toward a History of Image-Processed Video - Steina and Woody Vasulka," *Afterimage*, December 1983, p. 14.

400 Ibid., p. 48.

401 Steina Vasulka, "Reflections," *Steina e Woody Vasulka: Video, media e nuove immagini nell' arte contemporanea*, Palazzo delle Esposizioni, Rome 1995, p. 98 (Italian/English).

402 *Vidéographie*, 3. Semaine Internationale de Vidéo, Saint-Gervais, Maison des Jeunes et de la Culture, Geneva, Switzerland, 1989 , p. 43.

403 Lucinda Furlong, "Notes Toward a History of Image-Processed Video - Steina and Woody Vasulka," p. 15.

404 Charles Hagen, "An Interview with Woody Vasulka," *Afterimage*, Summer 1978, p. 20.

405 Scott Nygren, "Didactic Video: Organizational Models of the Electronic Image, Introduction," *Afterimage*, October 1975, p. 9.

406 Woody Vasulka, "A Syntax of Binary Codes," *Afterimage*, Summer 1978, p. 20.

407 Charles Hagen, "An Interview with Woody Vasulka," p. 26.

408 Ibid., p. 22.

409 Artist's statement. Quoted from: Marita Sturken, "In dialogue with the Machine," *Steina and Woody Vasulka. Machine Media*, San Francisco Museum of Modern Art, San Francisco 1996, p. 44.

410 Steina Vasulka, "Reflections," *Steina e Woody Vasulka: Video, media e nuove immagini nell' arte contemporanea*, p. 99.

411 Steina Vasulka, Artist's Statement, *American Landscape Video - The Electronic Grove*, The Carnegie Museum of Art, Pittburgh, PA 1988, p. 103.

412 Steina, interview with MaLin Wilson in the exhibition brochure *Scapes of Paradoxy: The Southwest and Iceland*, Jonson Gallery, University of New Mexico, Albuquerque 1986.

413 Marita Sturken, "Steina and Woody Vasulka: In Dialogue with the Machine," *Steina and Woody Vasulka. Machine Media*, p. 41.

414 Woody Vasulka and David Dunn, "The Theatre of Hybrid Automata," *ars electronica, Virtuelle Welten*, Band II, Linz, Austria, 1990, p. 266.

415 Ibid., p. 267.

416 Woody Vasulka, "From Printed Matter to Noncentric Space," *Steina and Woody Vasulka. Machine Media*, p. 70.

417 Woody Vasulka and David Dunn, "The Theatre of Hybrid Automata," *ars electronica, Virtuelle Welten*, Band II, p. 271.

418 Woody Vasulka, "From Printed Matter to Noncentric Space," *Steina and Woody Vasulka. Machine Media*, pp. 70-71.

419 Woody Vasulka and David Dunn, "The Theatre of Hybrid Automata," *ars electronica, Virtuelle Welten*, Band II, p. 271.

420 Shalom Gorowitz, in an interview with Lucinda Furlong, quoted in *Afterimage*, December 1983, p. 12.

421 Charles Hagen, "An Interview with Woody Vasulka," p. 21.

422 Ibid., p. 23.

423 David Dunn/Woody Vasulka, "Digital Space: A Research Proposal," a*rs electronica, Virtuelle Welten*, Band II, p. 270.

424 Similar trains of thought can be found in Gene Youngblood's media theories. "We're seeing a gradual evolution toward a synthesis of camera and synthetic imagery, the acquired and the generated image, as it were.." In: Peter Broderick, "Since Cinema Expanded: Interview with Gene Youngblood," *Millennium Film Journal*, 20th Anniversary Special Edition, Nos. 16, 17, 18, Fall/Winter 1986-1987, p. 60.

[425] Gene Youngblood traced the 'evolution' of cinematic language in the new technologies that came to the fore at the time, computer films, video/television, laser movies, multi-projection environments, in Marshall McLuhan's terminology, as developments toward "the global intermedia network." "The notion of expanded cinema is both technical, in terms of new instruments for generating the audio-visual stream, but also sociological. I'm suggesting that it's politically and socially more useful, more appropriate to think of cinema this way than to define it in medium-specific terms," thus Youngblood in an interview conducted by Peter Broderick, *Millennium Film Journal*, 1986-87, pp. 55-66.

[426] Stewart Brand, *Inventing the Future at MIT - The Media Lab*, pp. 235-239.

[427] Steven Durland, "defining the image as place - a conversation with kit galloway, sherrie rabinowitz & gene youngblood," *High Performance*, 37, 1987, p. 57.

[428] Statement, ibid., p. 58.

[429] Jonathan Benthall, *Science and Technology in Art Today*, Thames & Hudson, London 1972.

[430] Steward Brand, *Inventing the Future at MIT - The Media Lab*, p. 83.

[431] I must emphasize that this production process is different from the artist who hires assistants, like Sol LeWitt, Alice Aycock and many other sculptors do. Obviously, we have a different kind of collaborative process here.

[432] The term post-industrial society was first put forth in a paper by Daniel Bell, "The Post-Industrial Society: A Speculative View of the United States in 1985 and Beyond," written in 1959, but which remained unpublished. See Daniel Bell, *The Coming of Post-Industrial Society*, Basic Books, Inc., New York 1973, p. 36, n. 43. Daniel Bell's theorems about the changes from an industrial to a post-industrial society, in which the knowledge and information structure would be the predominant feature, were very influential in the United States.

[433] Raymond Bellour, "Interview with Bill Viola," *October*, 34, Fall 1985, pp. 112-113. Also: Nam June Paik, "Random Access Information," *Artforum*, September 1980, p. 46.

[434] Howard Rheingold, *The Virtual Community*, cited from www.well.com/user/hrl/vcbook.

[435] Robert Atkins, "The Art World & I Go On Line," *Art in America*, December 1995, pp. 58-65.

BIBLIOGRAPHY

Selected further reading is marked with an *

Periodicals

Art and Ecology, *Art Journal*, Summer 1992

Art and Social Consciousness, *Leonardo*, Vol. 26, No. 5, 1993

Ästhetik des Immateriellen? Teil I, *Kunstforum International*, Band 97, Nov./Dec. 1988

Ästhetik des Immateriellen? Teil II, *Kunstforum International*, Band 98, Jan./Feb. 1989

Computer Art in Context, SIGGRAPH 1989, *Leonardo*, Supplemental Issue, 1989

Digital Image/Digital Cinema, SIGGRAPH 1990, *Leonardo*, Supplemental Issue, 1990

E.A.T. News, January 15, 1967, until 1968

E.A.T. Operations and Information, 1968-1969

E.A.T. Proceedings, No. 1, 1969 - No. 13, 1977

Electronic Art, *Leonardo*, Supplemental Issue, 1988

Die globale Revolution, Bericht des Club of Rome, Spiegel-Spezial, Nr.1, Hamburg 1991

Im Netz der Systeme: Für eine interaktive Kunst: Ars Electronica, *Kunstforum International*, Band 103, Sept./Oct. 1989

Leonardo, 1968 -

Radical Software, Raindance Foundation, 1970-1974

TECHNE (A Projects and Process Paper), 1969-1970

Wired, November 1993 -

Catalogues

3ème *Semaine Internationale de Vidéo*. Maison des Jeunes et de la Culture, Saint-Gervais, Genève, 1989

Alan Sonfist 1969-1989. Brookville, NY: Hillwood Art Museum / Long Island University, 1989

Alaskan Impressions. Anchorage AL: The Visual Arts Center of Alaska, 1986

American Landscape Video - The Electronic Grove. Pittsburgh: The Carnegie Museum of Art, 1988

Andre, Buren, Irwin, Nordman: Space as Support. Berkeley, CA: University Art Museum, 1980

Arroyo Seco Release: A Proposal for Pasadena. Pasadena, CA: Baxter Art Gallery and California Institute of Technology, 1985

Atempause für den Save Fluss: Die Summe seiner Geschichte, Beginn einer neuen Geschichte. Helen Mayer Harrison and Newton Harrison. Berlin: Neuer Berliner Kunstverein in Zusammenarbeit mit dem Berliner Künstlerprogramm des D.A.A.D., 1989

Artpark (1974-). Lewiston, NY: The Program in the Visual Arts, 1974

ARTTRANSITION. Cambridge, MA: Center for Advanced Visual Studies/MIT, 1975

Art for 25 Million People. Berlin: Deutscher Akademischer Austausch Dienst, 1984

The Artist as Social Designer. Los Angeles: Los Angeles County Museum of Art, 1985

Bill Viola. Paris: ARC Musée d'Art de la Ville de Paris, 1983

Bill Viola. New York: Museum of Modern Art, 1987

Bill Viola: A Survey of a Decade. ed. Marilyn Zeitlin, Houston: Contemporary Arts Museum, 1988

Bill Viola: Unseen Images. ed. Marie Luise Syring, Düsseldorf: Städtische Kunsthalle, Verlag R. Meyer, 1992

Bill Viola. ed. Alexander Pühringer, Salzburg: Salzburger Kunstverein, Verlag Ritter Klagenfurt, 1994

Bill Viola: Buried Secrets. Tempe, AZ: Arizona State University Art Museum, in association with the Kestner-Gesellschaft, Hannover 1995

Boston Celebrations. March 18 - April 29, Cambridge, MA: MIT Press, 1975
California Perceptions: Light and Space. Selections from the Wortz Collection. Fullerton: California State University, Fullerton Art Gallery, 1979
Common Ground: Five Artists in the Florida Landscape. Sarasota, FL: The John and Mable Ringling Museum of Art, 1982
Devil's Gate Transformation: A Refuge for Pasadena. Pasadena, CA: Pasadena Gallery of Contemporary Art and Art Center College of Design, 1987
Earth Art. Ithaca, NY: Andrew Dickson White Museum of Art, Cornell University, 1969
Earthworks: Land Reclamation as Sculpture. Seattle, WA: Seattle Art Museum, 1979
Earthworks: Land Reclamation as Sculpture - Technical Report, King County Arts Commission, Seattle, WA, 1981
Elements: Earth, Air, Fire, Water. Boston, MA: Boston Museum of Fine Arts, 1971
Explorations. Washington D.C.: National Collection of Fine Arts, Smithsonian Institution, 1970
Fragile Ecologies: Contemporary Artists' Interpretations and Solutions. ed. Barbara C. Matilsky. New York: Rizzoli International Publications in association with the Queens Museum of Art, 1993
Framing and Being Framed, 7 works 70-75. Halifax: Nova Scotia College of Art and Design, 1975
Future Garden. Teil I. Die gefährdeten Wiesen Europas. Helen Mayer Harrison, Newton Harrison und Harrison Studio. Bonn: Kunst- und Ausstellungshalle der Bundesrepublik Deutschland, 1996
Grand Rapids Project: Robert Morris. Grand Rapids Art Museum, Grand Rapids, MI, 1975
Green Heart Vision. Newton Harrison, Helen Mayer Harrison and The Harrison Studio with The Green Heart Harrison Studio, Del Mar: The Harrison Studio, in association with the Cultural Council and Province of South Holland, Den Haag, Netherlands, 1995
Guadelupe Meander - A Refugia for San Jose. Helen and Newton Harrison.San Jose, CA: San Jose State University, 1983
György Kepes: MIT Years 1945-1977. Cambridge, MA: MIT Press, 1978
Hans Haacke: Volume I. Oxford/Eindhoven: Museum of Modern Art/Stedelijk van Abbemuseum, 1984
Hans Haacke: Vol. II, Works 1978-1983. Tate Gallery/Stedelijk van Abbemuseum, London/ Eindhoven, 1984
Helen Mayer Harrison and Newton Harrison - Baltimore Promenade. Baltimore, MD: Maryland Institute/Baltimore College of Art, 1981
James Turrell: Light and Space. Whitney Museum of American Art, New York, 1981
Jim Turrell, Stedelijk Museum, Amsterdam, 1976
Light, Motion, Space. Minneapolis: Walker Art Center, 1967
Lloyd Hamrol: Works, Projects, Proposals. Los Angeles: Municipal Art Gallery, 1986
Multiple Interaction Team. Cambridge, MA: Center for Advanced Visual Studies/MIT, 1972
Nam June Paik: Videa 'n' Videology (1959-1973). Syracuse, NY: Everson Museum of Art, 1974
Nam June Paik, Werke 1946-1976, Musik Fluxus Video. Cologne: Kölnischer Kunstverein, 1976
Nam June Paik: Video Time - Video Space. Basel/Zürich: Kunsthalle Basel, Kunsthaus Zürich, 1991
New American Landscape. Poughkeepsie, NY: Vassar College Art Museum, 1973
Nine Evenings: Theatre and Engineering. New York: Foundation for the Performing Arts, 1966
Occluded Front: James Turrell. Los Angeles: The Museum of Contemporary Art/The Lapis Press, 1985
Otto Piene und das CAVS. Karlsruhe: Badischer Kunstverein, 1988
Patricia Johanson: Public Landscapes. Philadelphia: Painted Bride Art Center, 1991
Robert Irwin. Chicago: Museum of Contemporary Art, 1975
Robert Irwin. New York: Whitney Museum of American Art, 1977
Robert Irwin. New York: Rizzoli International Publications, in association with The Museum of Contemporary Art, Los Angeles, 1993
The Lagoon Cycle. Ithaca and New York: Herbert F.Johnson Museum / Cornell University 1985
The Last Time I saw Ferus 1957-1966. Newport Beach: Newport Harbor Art Museum, 1976

The Machine as Seen at the End of the Mechanical Age. ed. K.G. Pontus Hultén, New York: Museum of Modern Art, 1968

The Serpentine Lattice. Newton Harrison and Helen Mayer Harrison. Portland, OR and New York: The Douglas F. Cooley Memorial Art Gallery, Reed College and Ronald Feldman Fine Arts, 1993

SKYART Conference '81. Cambridge, MA: Center for Advanced Visual Studies/MIT, 1981

Software, Information Technology: Its Meaning for Art, New York: Jewish Museum, 1970

Some More Beginnings. Brooklyn, NY: Brooklyn Museum of Art, 1968

Sonsbeek. Arnhem, Netherlands: Sonsbeek Foundation, 1971

Steina e Woody Vasulka. Video, media e nuove immagini nell' arte contemporanea. Rome: Palazzo delle Esposizioni, 1995

Steina and Woody Vasulka. Machine Media. San Francisco: San Francisco Museum of Modern Art, 1996

TV as a Creative Medium. New York: Howard Wise Gallery, 1969

Vasulka. Steina: machine vision/Woody: descriptions. Buffalo, NY: Albright-Knox Art Gallery, 1978

Vision and Television. Foreword by Russel Connor, Waltham, MA: Rose Art Museum, Brandeis University, 1970

Articles and Books

Adcock, Graig. "Perceptual Edges: The Psychology of James Turrell's Light and Space," *Arts Magazine*, Feb. 1985, pp. 124-128

Adcock, Graig. "Time: The Visual Parameters of the Roden Crater," *Occluded Front: James Turrell*, Los Angeles: The Museum of Contemporary Art/The Lapis Press,1985, pp. 101-135

Adcock, Graig. James Turrell: *The Art of Light and Space*, Berkeley/Los Angeles: University of California Press, 1992

Adcock, Graig. "Conversational Drift. Interview with Helen Mayer Harrison and Newton Harrison," *Art Journal*, Summer 1992, pp. 35-45

Allen, Jerry. "How Art Becomes Public," *Going Public: A Field Guide to Developments in Art in Public Places*, Amherst, MA: Arts Extension Service/University of Massachusetts, 1988, pp. 244-251

Alloway, Lawrence, "Systemic Painting," *Systemic Painting*, New York: Solomon R. Guggenheim Foundation, 1966

Alloway, Lawrence. "Robert Smithson's Development," *Artforum*, Nov. 1972, pp. 53-60

Alloway, Lawrence. "Art," *Nation* 219, December 21, 1974, p. 670

Alloway, Lawrence. *An Autobiography of Alan Sonfist*, Ithaca, NY: Herbert F. Johnson Museum of Art, Cornell University, 1975

Alloway, Lawrence, Mark Roskill and Nicolas Calas. "Systems of Cross-References in the Arts," *Arts Magazine*, Summer 1971, p. 16

Antin, David. "Television: Video's Frightful Parent," *Artforum*, Dec. 1975, pp. 36-45

Apel, Karl Otto. *Charles S. Peirce - From Pragmatism to Pragmaticism*, Amherst, MA: University of Massachusetts Press, 1981

Apter, Michael. "Cybernetics and Art," *Leonardo*, Vol. 2, No. 33, July 1969, pp. 257-265

Ashby, W. Ross. *An Introduction to Cybernetics*, New York: John Wiley & Sons, 1958

Ashton, Dore. "Exercises in Anti-Style," *Arts Magazine*, April 1969, pp. 45-47

Atkins, Robert. "The Art World & I Go On Line," *Art in America*, Dec. 1995, pp. 58-65

Auping, Michael. "Interview with Alan Sonfist," *Common Ground: Five Artists in the Florida Landscape*, Sarasota, FL: The John and Mable Ringling Museum of Art, 1982, pp. 133-136

Auping, Michael. "Interview with Helen and Newton Harrison," *Common Ground: Five Artists in the Florida Landscape*, Sarasota, FL: The John and Mable Ringling Museum of Art, 1982, pp 98-104

Avgikos, G. "Greenpiece," *Artforum*, April 1991, pp. 104-110

*Bachelard, Gaston. *The Poetics of Space*, Boston: Beacon Press, 1969

Balken, Debra Bricker. "Robert Irwin at Pace," *Art in America*, Jan. 1993, p. 99

Bateson, Gregory. "The Cybernetics of 'Self': A Theory of Alcoholism," *Psychiatry*, Vol. 34, No. 1, 1971, pp. 1-18

Bateson, Gregory. *Steps to an Ecology of Mind*, New York: Ballantine/Psychology, 1972

Bateson, Gregory. "Minimal Requirements for a Theory of Schizophrenia," *Steps to an Ecology of Mind*, New York: Ballantine/Psychology, 1972, pp. 244-270

Battcock, Gregory. ed., *Minimal Art: A Critical Anthology*, New York: E.P. Dutton, 1968

Beardsley, John. *Probing the Earth: Contemporary Land Projects*, Washington D.C.: Hirshhorn Museum and Sculpture Garden, Smithsonian Institution, 1977

Beardsley, John. *Earthworks and Beyond*, New York: Abbeville Press, 1984

Beardsley, John. ed., *Art in Public Places: A Survey of Community Sponsored Projects*, Washington D.C.: Partners for Livable Spaces, 1981

Becker, Howard S./Walton, John. eds., "Social Science and the Work of Hans Haacke," *Framing and Being Framed. 7 Works '70-'75*, Halifax: Nova Scotia College of Art and Design, 1975, pp 145-152

*Benedikt, Michael. ed., *Cyberspace: First Steps*, Cambridge, MA: MIT Press, 1991

Belloli, Jay. *Arroyo Seco Release/A Serpentine for Pasadena*, Pasadena: Baxter Art Gallery, California Institute of Technology, 1985

Bellour, Raymond. "An Interview with Bill Viola," *October*, No. 34, Fall 1985, pp. 91-119

*Bellour, Raymond/Duguet, Anne-Marie. eds. ,*Vidéo, Communications*, Vol. 48, Paris: Ed. du Seuil, 1988

Bell, Daniel. *The Coming of Post-Industrial Society*, New York: Basic Books, 1973.

Benthall, Jonathan. *Science and Technology in Art Today*, London: Thames and Hudson, 1972

Benthall, Jonathan. "The Prehistory of the Invisible," *Studio International*, Nov. 1969, p. 152

Benthall, Jonathan. "Haacke, Sonfist, and Nature," *Studio International*, March 1971, pp. 95-96

Benthall, Jonathan. "Newton Harrison: big fish in a small pool," *Studio International*, Dec. 1971, pp. 132-143

Benthall, Jonathan, "Kepes' Center at MIT," *Art International*, Vol. 19, No. 1, Jan. 1975, pp. 28-31+49

Berman, Avis. "Public Sculpture's New Look," *Art News*, September 1991, pp. 102-109

Berger, René. "Video and the Restructuring of Myth," *The New Television: A Public/Private Eye*, Cambridge/London: MIT Press, 1978

*Berman, Marshall. *All That is Sold Melts Into Air: Modernism and Modernization*, New York: Simon & Schuster, 1983

Bertalanffy, Ludwig von, *General System Theory*, New York: George Braziller, 1968

Bertalanffy, Ludwig von, *Perspectives in General System Theory*, New York: George Braziller, 1974

Bienstock, David. *A Special Videotape Show*, program notes, New York: Whitney Museum of American Art, 1971

Bijvoet, Marga. "In Search of Self Through Nature: Art Experiments in the California Desert," *Arts and Architecture*, Winter 1983, pp. 40-43

Bijvoet, Marga. "How Intimate Can Art and Technology Really Be?" *Culture, Technology, and Creativity*, ed. Philip Hayward, London/Paris: John Libbey, 1988, pp. 15-37

Birmingham, Peter. *Alan Sonfist/Trees*, Washington D.C: Smithsonian Institution, 1978.

*Bohm, David. *Wholeness and the Implicate Order*, London: ARK pb., 1950.

Bois, Yves-Alain, Douglas Crimp and Rosalind Krauss. "A Conversation With Hans Haacke," *October*, No. 30, Fall 1984, pp. 49-81

Bongartz, Roy. "It's Called Earth Art - And Boulderdash," *New York Times Magazine*, February 1, 1970, pp. 16-17, pp. 22-30

Boyle, Deirdre. "Interview with Bill Viola," *Bill Viola*, Paris: ARC Musée d'Art Moderne de la Ville de Paris, 1983, p. 14

Brand, Stewart. *The Media Lab: Inventing the Future at MIT*, Harmondsworth: Penguin Books, 1987

Broderick, Peter. "Since Cinema Expanded: Interview With Gene Youngblood," *Millenium Film Journal*, Nos.16/17/18, Fall/Winter 1986-87, pp. 55-66

Buchloh, Benjamin H.D. "Hans Haacke: Memory and Instrumental Reason," *Art in America*, Feb. 1988, pp. 97-107, pp.157-159

Burgess, Lowry. "Art and Science Project: 'Waiting-for-Light-Planes'," *Leonardo*, Vol. 7, No. 4, 1974, pp. 329-331

Burnham, Jack."Hans Haacke: Wind and Water Sculpture," *Tri-quarterly Supplement*, No. 1, Spring 1967, pp. 1-24

Burnham, Jack. *Beyond Modern Sculpture*, New York: George Braziller, 1968

Burnham, Jack. "Systems Esthetics," *Artforum*, Sept., 1968, pp. 30-35

Burnham, Jack. "Real Time Systems," *Artforum*, Sept. 1969, pp. 49-55

Burnham, Jack. The Structure of Art, New York: George Braziller, 1970

Burnham, Jack. "The Aesthetics of Intelligent Systems," *On the Future of Art*, New York: Viking Press, 1970, pp. 95-122

Burnham, Jack. "Notes on Art and Information Processing," *Software, Information Technology: Its New Meaning for Art*, New York: Jewish Museum, 1970, pp. 10-14

Burnham, Jack. "Corporate Art," *Artforum*, Oct. 1971, pp. 68-71

*Burnham, Jack. *Great Western Salt Works*, New York: George Braziller, 1974

Burnham, Jack. "Art and Technology: The Panacea that Failed," *Video Culture: A Critical Investigation*, ed. John G. Hanhardt, Rochester, NY: Visual Studies Workshop Press, 1986, pp. 232-248

Burnham, Linda Frye. "Between the Diaspora and the Crinoline," An Interview with Bonnie Sherk, *High Performance*, Fall 1981, pp. 49-51

Butterfield, Jan. "The State of the Real: Robert Irwin Discusses the Activities of an Extended Consciousness," *Arts Magazine*, June 1972, pp. 47-49

Butterfield, Jan. "Robert Irwin: On the Periphery of Knowing," *Arts Magazine*, Feb. 1976, pp. 72-77

Butterfield, Jan. *The Art of Light and Space*, New York: Abbeville Press, 1993.

Brigham, Joan. "Reclamation Artists, " *Leonardo*, Vol. 26, No. 5, 1993, pp. 387-395

Cage, John. *Silence*, Middletown, CT: Wesleyan University Press, 1961

Campbell, Robert, and Jeffrey Cruikshank. "Art in Architecture," *Designing the Wiesner Building*, Cambridge, MA: MIT Press, (n.d.), pp. 11-24

Carsen, Rachel. *Silent Spring*, Boston: Houghton Mifflin, 1962

Cembalist, R. "The Ecological Art Explosion," *Art News*, Summer 1991, pp. 96-105

Chandler, John Noel. "Hans Haacke: The Continuity of Change," *Artscanada*, June 1969, pp. 8-11

Cohen, Harold. "On Purpose: An Enquiry Into the Possible Roles of the Computer in Art," *Studio International*, Jan. 1974, pp. 9-16

Cruikshank, Jeffrey L. and Pam Korza. eds., *Going Public: A field guide to developments in art in public places*, Amherst, MA: Arts Extension Service/Visual Arts Program of the National Endowment for the Arts, 1988

Cummings, Paul. "Interview with Robert Smithson for the Archives of American Art," *The Writings of Robert Smithson*, ed. Nancy Holt. New York: New York University Press, 1979, pp. 137-156

Daniels, Dieter. "Bill Viola: Installations and Videotapes," *Kunstforum International*, No. 92, December 1987/January 1988, pp. 247-250

Davis, Douglas M. "Art and Technology - The New Combine," *Art in America*, Jan./Feb. 1968, pp. 28-47

Davis, Douglas. *Art and the Future: A History/Prophecy of the Collaboration between Science, Technology and Art*, New York: Praeger, 1973

Davis, Douglas. "Robert Rauschenberg: Technology as Nature," *Art and the Future*, New York: Praeger, 1973, pp. 142-145

Davis, Douglas. "Nam June Paik: The Cathode Ray Canvas," *Art and the Future*, New York: Praeger, 1973, pp. 146-154

Davis, Douglas M. "Conversations with György Kepes, Billy Klüver and James Seawright, " *Art in America*, Jan./Feb. 1968, pp. 38-45

Davis, Douglas M. "Art and Technology - The New Combine," *Art in America*, Jan./Feb. 1968, pp. 28-37

Davis, Douglas and Allison Simmons. eds. *The New Television: A Public/Private Eye*, Cambridge, MA/London: MIT Press, 1978.

Decker, Edith. *Paik Video*, Cologne: DuMont, 1988

Deitsch, Jeffrey. "The New Economics of Environmental Art," *Art in the Land*, ed. Alan Sonfist, New York: E.P. Dutton, 1983, pp. 85-91

Denes, Agnes. "Notes on Eco-logic Environmental Artwork, Visual Philosophy and Global Perspective," *Leonardo*, Vol. 26, No. 5, 1993, pp. 387-395

Denes, Agnes. *The Beginning and the End and Thereafter*, Rochester, NY: Visual Studies Workshop, 1989.

Durland, Steve. "Bill Viola: Passages," *High Performance*, Vol. 20, No. 40, 1988, p. 84

Drucker, Peter. *The Age of Discontinuity*, London: Heinemann, 1969

Dunn, David, Woody Vasulka. "The Theatre of Hybrid Automata," *ars electronica, Virtuelle Welten*, Band II, Linz, 1990, pp. 265-267

Dunn, David, Woody Vasulka. "Digital Space: A Research Proposal," *ars electronica, Virtuelle Welten*, Band II, Linz, 1990, pp. 268-275

Durland, Steven. "defining the image as place - a conversation with kit galloway, sherry rabinowitz, and gene youngblood" *High Performance*, Issue 37, 1987, pp. 57-58

Einstein, Albert. *The World As I See It*, New York: Citadel Press, 1979 (1st ed. 1934)

*Ellul, Jacques. *The Technological Society*, New York: Random House/Vintage Books, 1964

Esterow, Milton. "How Public Art Becomes a Political Hot Potato," *Art News*, Jan. 1986, pp. 75-79

Felshin, Nina. *But Is It Art? The Spirit of Art as Activism*, Seattle: Bay Press, 1995

*Flusser, Vilém. *Ins Universum der technischen Bilder*, Göttingen: European Photography, 1985.

Fried, Michael. "Art and Objecthood," *Artforum*, Summer 1967, pp. 12-33

Fry, Edward F. *Hans Haacke: Werkmonographie*, Cologne: DuMont Kunst-Praxis, 1972.

Fuller, Buckminster R. *Operating Manual for Spaceship Earth*, Carbondale, IL: Southern Illinois University Press, 1969

Furlong, Lucinda. "Notes Toward a History of Image-Processed Video: Steina and Woody Vasulka," *Afterimage*, Dec. 1983, pp. 12-17

Furlong, Lucinda. "Getting High Tech: The New Television," *The Independent*, Vol. 8, No. 2, March 1985

Gablik, S. "Toward an Ecological Self," *New Art Examiner*, Jan. 1992, pp. 26-30

Gablik, S. "The Ecological Imparative," *Art Journal*, Summer 1992, pp. 49-51

Gill, Johanna. *Video: State-of-the-Art*, New York: Rockefeller Foundation, 1976

Glueck, Grace. "Nature's Boy," *The New York Times*, Nov. 15, 1970

Glueck, Grace. "Art people: The Earth is Their Palette," *The New York Times*, April 4, 1980, sect. C, p. 24

Goodman, Cynthia. *Digital Visions*, New York/Syracuse: Harry N. Abrams/Everson Museum of Art, 1987.

Gribbin, John. *In Search of Schrödinger's Cat - Quantum Physics and Reality*, London: Wildwood House, 1984

Gruen, John. "Art Meets Technology," *World Journal Tribune*, Magazine Section, New York, October 2, 1966. In: *E.A.T Clippings*, Vol. 1, No. 1, pp. 4-5

Haacke, Hans. "Visitors' Profile: A statistical breakdown of spectators," *Software, Information Technology: Its New Meaning for Art*, New York: Jewish Museum, 1970, p. 34

Haacke, Hans. "John Weber Gallery Visitors' Profile 1," exhibition brochure, New York: John Weber Gallery, 1972 (n.p.)

Haacke, Hans. "John Weber Gallery Visitors' Profile 2," exhibition brochure, New York: John Weber Gallery, 1973 (n.p.)

Haacke, Hans and Bruce Boice. "Hans Haacke's Gallery Visitors' Profile," *Artforum*, June 1973, pp. 44-46

Haacke, Hans. "Museums, Managers, and Consciousness," *Hans Haacke: Unfinished Business*, New York/Cambridge, MA: The New Museum of Contemporary Art/MIT Press, 1986, pp. 33-40

Hagen, Charles. "Interview with Woody Vasulka," *Afterimage*, Summer 1985, pp. 20-31

Hanhardt, John G. ed., *Video Culture: A Critical Investigation*, Rochester, NY: Visual Studies Workshop Press, 1986

Hanhardt, John G. "Video Art: Expanded Forms," *Leonardo*, Vol. 23, No. 4, 1990, pp. 437-439

Harithas, James. ed., "Introduction," *Nam June Paik: Videa 'n'Videology*, Syracuse, NY: Everson Museum of Art, 1974, (n.p.)

Harrison, Newton. "Sea Grant second narrative and two precedent works," *Studio International*, May 1974, pp. 234-237

Harrison, Helen Mayer and Newton Harrison. "Spoils File Reclamation Project," *Artpark*, Lewiston, NY: The Program in the Visual Arts, 1977, pp. 40-43

Harrison, Helen Mayer and Newton Harrison. *The Lagoon Cycle*, Ithaca, NY: Herbert F. Johnson Museum of Art, Cornell University, 1985

Harrison, Helen Mayer and Newton Harrison. "Shifting Positions Toward the Earth: Art and Environmental Awareness," *Leonardo*, Vol. 26, No. 5, 1993, pp. 371-387

Harrison, Newton and Helen Mayer Harrison. "Tibet is the High Ground," *Environmental Action*, July/Aug. 1991

Harrison, Newton and Helen Mayer Harrison. "A Lattice or a Serpentine: The Beginning of a Conceptual Design," *Seattle Arts*, Seattle Arts Commission, Vol. 15, No. 10, 1991

Harris, Stacy Paleologus. ed., *Insights/On Sites: Perspectives on Art in Public Places*, Washington D.C.: Partners for Livable Places, 1984

Hawking, Stephen W. *A Brief History of Time*, New York: Bantam Books, 1988

*Hayward, Philip. ed., *Picture This: Media Representations of Visual Art & Artists*, London/Paris: John Libbey, 1988

Hayward, Philip. ed., Culture, *Technology & Creativity in the Late Twentieth Century*, London/Paris/Rome: John Libbey, 1990

*Heim, Michael. *The Metaphysics of Virtual Reality*, New York: Oxford University Press, 1994

Herzogenrath, Wulf and Edith Decker. eds., *Video Skulptur, retrospektiv und aktuell*, Cologne: DuMont, 1987

Hess, A. "Technology exposed toward a new aesthetic for technology in the landscape," *Landscape Architecture*, May 1992, pp. 44-45

Hess, Thomas B. "Gerbils ex Machina," *Art News*, Dec. 1970, p. 73

Hobbs, Robert. *Robert Smithson: Sculpture*, Ithaca, NY/London: Cornell University Press, 1981

Hoberman, J. "The Reflecting Pool: Bill Viola and the Visionary Company," *Bill Viola*, New York: Museum of Modern Art, 1987, pp. 63-72

Holt, Nancy. "Hydra's Head," *Arts Magazine*, Jan. 1975, pp. 57-59

Holt, Nancy. "Sun Tunnels," *Artforum*, April 1977, pp. 32-37

Holt, Nancy. "Stone Enclosure: Rock Rings, 1977-78," *Arts Magazine*, June 1979, pp. 152-155

Holt, Nancy. *Ransacked*, New York: Printed Matter, 1980

Holt, Nancy. "Pipeline," *Alaskan Impressions*, Anchorage, AL: The Visual Arts Center of Alaska, 1986

Horwitz, Robert Joseph. "Nature as Artifact: Alan Sonfist," *Artforum*, Nov. 1973, pp. 32-35

Huffman, Kathy, R. ed., *Video: A Retrospective 1974-1984*, Long Beach, CA: Long Beach Museum of Art, 1984

Irwin, Robert. "Notes Toward a Model," *Robert Irwin*, New York: Whitney Museum of American Art, 1977, pp. 23-31

Irwin, Robert. *Being and Circumstance: Notes Toward a Conditional Art*, Larkspur Landing, CA: The Lapis Press in conjunction with the San Francisco Museum of Modern Art and The Pace Gallery, 1985

Irwin, Robert, "The Hidden Structures of Art, " *Robert Irwin*, New York/Los Angeles: Rizzoli International Publications and The Museum of Contemporary Art, 1993, pp. 13-49
Iskin, Ruth. "The Fine Art of Social Change," *Social Works*, Los Angeles: Los Angeles Institute of Contemporary Art, 1979, pp. 8-16
Jameson, Fredric. "Hans Haacke and the Cultural Logic of Postmodernism," *Hans Haacke: Unfinished Business*, New York/Cambridge, MA: The New Museum of Contemporary Art/MIT Press, 1986, pp. 42-43
Johnson, Ken. "Poetry and Public Service," *Art in America*, March 1990, pp. 161-163+219
Johnson, Ken. "Alan Sonfist at Ledisflam," *Art in America*, Jan. 1992, p. 109
Kay, Jane H. "Art and Science on the Charles," *Art in America*, Summer 1967, pp. 62-67
Kepes, György. *The Language of Vision*, Chicago: Paul Theobald, 1944
Kepes, György. ed., *The New Landscape in Art and Science*, Cambridge, MA: MIT Press, 1956
*Kepes, György. ed., *Structure in Art and Science*, VISION+VALUE Series, New York: George Braziller, 1965
Kepes, György. "The Visual Arts and the Sciences: A Proposal for Collaboration," *Daedalus*, Vol. 94, No. 1, 1965, pp. 117-133
Kepes, György. "The Lost Pageantry of Nature," *Artscanada*, Dec. 1968, pp. 33-39
Kepes, György. "Toward Civic Art," *Leonardo*, Vol. 4, No. 1, Winter 1971, pp. 69-73
Kepes, György.ed., *Arts of the Environment*, New York: George Braziller, 1972
Kepes, György. "Art and Ecological Consciousness," *Arts of the Environment*, pp. 1-12
Kepes, György. "The Artist's Role in Environmental Self-regulation," *Arts of the Environment*, pp. 167-191
Kern, Hermann. "Labyrinths: Tradition and Contemporary Works," *Artforum*, May 1981, pp. 60-68
Klüver, Billy. "9 Evenings: Theatre and Engineering," *Nine Evenings: Theatre and Engineering*. New York: Foundation for the Performing Arts, 1966, (n.p.)
Klüver, Billy. "Remarks by billy Klüver," *EAT News*, Vol. 1, No. 3, November 1, 1967, (n.p.)
Klüver, Billy. "Art and Science: Two Worlds Meet," *Bell Telephone Magazine*, Nov./Dec. 1967
Klüver, Billy. "The Pavilion," *Pavilion*, ed. Billy Klüver/Julie Martin/Barbara Rose, New York: E.P. Dutton, 1972, pp. ix-xvi
Klüver, Billy, "The Artist's Expertise For Communication Planning," *EAT Proceedings*, No. 13, December 15, 1977, (n.p.)
Krauss, Rosalind. "Sculpture in the Expanded Field," *October*, No. 8, 1979, pp. 31-44
Kubler, George. *The Shape of Time. Remarks on the History of Things*, New Haven: Yale University Press, 1962
Kuhn, Thomas S. *The Structure of Scientific Revolutions*, New Haven/London/Chicago: University of Chicago Press, 1962
Kurtz, Bruce. "Paikvision," *Artforum*, Feb. 1982, pp. 52-55
Lacy, Suzanne. ed., *Mapping the Terrain: New Genre Public Art*, Seattle: Bay Press, 1995
Larson, Kay. "New Landscapes in Art," *New York Times Magazine*, May 13, 1979, pp. 20-23, 28-38
Larson, Kay. "Watering Wholes," *The Village Voice*, April 14, 1980, p. 78
Laszlo, Ervin. *Introduction to Systems Philosophy*, New York: Gordon and Breach, 1972
Leccese, M. "A Park with an Emptiness Below," *Landscape Architecture*, April 1993, pp. 52-58
Leider, Philip. "Robert Irwin," *Robert Irwin/Kenneth Price*, Los Angeles: Los Angeles County Museum of Art, 1966, (n.p.)
*Leveson, Paul. *A Sense of the Earth*, New York: Anchor Natural History Books/Doubleday, 1972
LeVeque, Terry Ryan. "Nancy Holt's 'Sky Mound': Adaptive Technology Creates Celestial Perspectives," *Landscape Architecture*, April/May 1988, pp. 82-86
Levin, Kim. "Helen and Newton Harrison: New Grounds for Art," *Arts Magazine*, Feb. 1978, pp. 126-129
Levine, Edward. "Robert Irwin's Recent Work," *Artforum*, Dec. 1977, pp. 24-29
Licht, Ira. "Robert Irwin," *Robert Irwin*, Chicago: Museum of Contemporary Art, 1975, (n.p.)

Lindgren, Nilo. "Art and Technology I - Steps Toward a New Synergism," *IEEE Spectrum*, April 1969, pp. 59-68

Lindgren, Nilo. "Art and Technology II - A Call for Collaboration," *IEEE Spectrum*, May 1969, pp. 46-56

Lindgren, Nilo. "Into the Collaboration," *Pavilion*, ed. Billy Klüver/Julie Martin/Barbara Rose, New York: E.P. Dutton, 1972

Lipke, William C. "Earth Systems," *Earth Art*, Ithaca, NY: Andrew Dickson White Museum of Art/ Cornell University, 1969, (n.p.)

Lipke, William C. "Fragments of a Conversation," *The Writings of Robert Smithson*, ed. Nancy Holt, New York: New York University Press, 1979, pp. 168-170

Lippard, Lucy R. "Total Theatre," *Art International*, Jan. 1967, pp. 39-40.

Lippard, Lucy R. *The Dematerialization of the Art Object*, New York: Praeger, 1973

Lippard, Lucy R. "Gardens: Some Metaphors for a Public Art," *Art in America*, October 1981, pp. 136-150

Lippard, Lucy R. *Overlay: Contemporary Art and the Art of Prehistory*, New York: Pantheon Books, 1983

Lippard, Lucy R. and John Chandler. "Visible Art and the Invisible World," *Art International*, Vol. 11, No. 5, May 1967, pp. 27-30

Lippard, Lucy R. and John Chandler. "The Dematerialization of Art," *Art International*, February 1968, pp. 31-36

Livingston, Jane. "Thoughts on Art and Technology," *A Report on the Art and Technology Program*, Los Angeles: Los Angeles County Museum of Art, 1971, pp. 43-47

*Lovejoy, Margot. *Postmodern Currents: Art and Artists in the Age of Electronic Media*, Ann Arbor: UMI Research Press, 1989

Lovelace, Cary. "Nancy Holt Brings the Heavens Down to Earth," *Arts*, New Jersey Council on the Arts, Fall 1987, pp. 10-14

Lovelock, James. *Gaia: A New Look at Life on Earth*, New York: Oxford University Press, 1987 (new edition, 1995).

Luhmann, Niklas. *Soziale Systeme*, Frankfurt, 1984

Mallory, Robert. "Computer Sculpture: Six Levels of Cybernetics," *Artforum*, May 1969

Mallory, Robert. "Notes on Jack Burnham's Concepts of a Software Exhibition," *Leonardo*, Vol. 3, No. 2, 1970, pp. 189-190

Mannheim, Karl. *Ideologie und Utopie*, Frankfurt/M.: Klostermann Verlag, 1985

Mathews, W.H. *Mazes and Labyrinths*, New York: Dover, 1970

Matzner, Florian. "A Short Trip on the Electronic Superhighway with Nam June Paik," *Nam June Paik: eine DATA base*, Cantz 1993, pp. 117-133

McCulloch, Warren. *Embodiment of Mind*, Cambridge, MA: MIT Press, 1965

McGill, Douglas C. "Architect and Artists Collaborate on Battery Park City Plaza," *New York Times*, January 31, 1985, III, p. 17.

McLuhan, Marshall. *The Gutenberg Galaxy*, New York: Signet Books, 1962

McLuhan, Marshall. *Understanding Media: The Extensions of Man*, New York: McGraw-Hill, 1964

McNamara, M.H. "Garden Technology (video arbor)," *Landscape Architecture*, March 1991, p. 26

Mekas, Jonas. "On New Directions, On Anti-Art, On the Old and the New in Art," *Movie Journal*, McMillan Co, New York 1972, p. 208

*Merchant, Carolyn. *The Death of Nature: Woman, Ecology and the Scientific Revolution*, San Francisco: Harper Collins, 1983

Miller, Jonathan. *McLuhan*, London: Collins, 1971

Morris, Robert. "Anti-Form," *Artforum*, April 1968, pp. 33-35Morris, Robert. "Notes on Art as/and Land Reclamation," *October*, No. 12, Spring 1980, pp. 87-102

Müller, Gregoire. "The Earth, subject to Cataclysms, is a Cruel Master - Interview with Robert Smithson," *Arts Magazine*, Nov. 1972

*Mumford, Lewis. *Myth of the Machine: The Pentagon of Power*, New York: Harcourt, 1970

Murray, Ann. "Pits and Pitfalls: Harriet Feigenbaum's Reclamation Art," *Woman's Art Journal*, Spring/Summer 1991, pp. 29-36

Nash, Michael. "Bill Viola's Re-Visions of Mortality," *High Performance*, Issue 37, 1987, pp. 63-64

Negroponte, Nicolas. *The Architecture Machine*, Cambridge, MA: MIT Press, 1970

Negroponte, Nicolas. *Being Digital*, New York: Alfred F. Knopff, 1994

Noll, Michael A. "The Digital Computer as a Creative Medium," *IEEE Spectrum*, Oct. 1967

Nygren, Scott. "Didactic Video: Organizational Models of the Electronic Image, Introduction," *Afterimage*, October 1975, pp. 9-13

*Oakes, Baile. ed., *Sculpting the Land - A Natural Dialog*, New York, 1995

Onorato, Ronald. "The Modern Maze," *Art International*, April/May 1976, pp. 21-25

Paik, Nam June. *Electronic Art*, brochure, New York: Galeria Bonino, 1965, (n.p.)

Paik, Nam June. "Video Synthesizer Plus," *Radical Software*, Vol. 1, No. 2, 1970. In cat. *Nam June Paik: Videa 'n' Videology (1959-1973)*, p. 59

Paik, Nam June. "Simulation of Human Eyes by 4-Channel Stereo Video Taping," *EAT/L.A. Survey*, No. 6, October 1970 In cat. *Nam June Paik: Videa 'n' Videology*, p. 62

Paik, Nam June. "Norbert Wiener and Marshall McLuhan," *Institute of Contemporary Arts Bulletin*, No. 3, London, 1972

Paik, Nam June. "Global Groove and Video Common Market," *The WNET-TV Lab News,* Issue #2, 1973. In cat. *Nam June Paik: Videa 'n' Videology*, pp. 60-61

Paik, Nam June. "Expanded Education for the Paperless Society," *Videa 'n' Videology*, Syracuse, NY: Everson Museum of Art, 1974. In cat. *Nam June Paik: Videa 'n' Videology*, pp. 31-39

Paik, Nam June. "Binghamton Letter," *Nam June Paik: Videa 'n' Videology*, pp. 68-69

Paik, Nam June. "Paper TV vs. Real TV," *Art Rite*, Vol. 7, 1974, p. 14

Paik, Nam June. "Medienplanung für das nachindustrielle Zeitalter (1974). Bis zum 21. Jahrhundert sind es nur noch 26 Jahre," *Nam June Paik, Werke 1946-1976, Musik Fluxus Video*, Cologne: Kölnischer Kunstverein, 1976, pp. 155-166

Paik, Nam June. "Random Access Information," *Artforum*, Dec. 1980, pp. 46-49

Paik, Nam June. *Niederschriften eines Kulturnomaden*, ed. E. Decker, Cologne: DuMont, 1992

Paik, Nam June. *Nam June Paik: eine DATA base*, eds. Klaus Bußmann, Florian Matzner, Cantz 1993

Peirce, Charles S. *Collected Papers*, Vols. I - VI, ed. C. Hawthorne and P. Weiss, Cambridge, MA: Harvard University Press, 1982

*Penrose, Roger. *The Emperor's New Mind*, New York: Oxford University Press, 1989

Perrault, John. "Long Live Earth," *The Village Voice*, October 17, 1968, p. 17

Perrault, John. "Down to Earth," *The Village Voice*, February 13, 1969, p. 18

Perrault, John. "To Make It New," *The Village Voice*, December 5, 1968, p. 16

Perrault, John. "Nonsites in the News," *The Village Voice*, February 24, 1969, pp. 44-46

Phillips, C. "Paik Video burns in Düsseldorf," *Art in America*, Nov. 1993, p. 31

Piene, Otto. "Arttransition," *ARTTRANSITION*, Cambridge, MA: Center for Advanced Visual Studies/MIT, 1975, (n.p.)

Plagemann, Volker. ed., *Kunst im öffentlichen Raum: Anstöße der 80er Jahre*, Cologne: DuMont Buchverlag, 1989

Platt, Susan. "Helen Mayer Harrison and Newton Harrison: An Urban Discourse," *Artweek*, May 21, 1983, pp. 1, 16

Popper, Frank. *Die kinetische Kunst*, Cologne: DuMont-Schauberg, 1975

Popper, Frank. *Art-action and participation*, New York: New York University Press, 1975

Preikschat, Wolfgang. *Video: Poesie der Neuen Medien*, Weinheim: Beltz Verlag, 1987

Preikschat, Wolfgang. "The Vasulkas - Vom Video-Feedback zur hybriden Interaktion," *Kunstforum International*, März 1992, pp. 184-193

*Price, Jonathan. *Video Visions. A Medium Discovers Itself*, New York: New American Library, 1977 (1972)

*Prigogine, Ilya and Isabelle Stengers. *Order out of Chaos*, New York: Bantam Books, 1984

Pulsa, "The City as Artwork," *Arts of the Environment*, ed. G. Kepes, New York: George Braziller, 1972, pp. 208-221

Ramljek, Suzanne. "Interview: Billy Klüver," *Sculpture*, May/June 1991, pp. 32-35

*Raskin, David. *Tracing Cultures - Art History, Criticism, Critical Fiction*, New York: Whitney Museum of American Art, 1995

*Raven, Arlene. *Art in the Public Interest*, Ann Arbor, MI: University of Michigan Press, 1988

Redstone, Louis G. and Ruth R. Redstone. *Public Art: New Directions*, New York: McGraw-Hill, 1981

Reich, Charles A. *The Greening of America*, New York: Random House, Inc., 1970

Reichardt, Jasia. *Cybernetic Serendipity: The Computer and the Arts*, London: Institute of Contemporary Art, 1968

Reichardt, Jasia. *Cybernetics, art and ideas*, Greenwich, CT/London: New York Graphic Society/Studio Vista, 1971

Rifkin, Jeremy. *Entropy: A New World View*, New York: Bantam Books/Viking Press, 1980

Rheingold, Howard. *The Virtual Community: Finding Connections in a Computerized World*, London: Minerva, 1995

Rheingold, Howard. *Virtual Reality: The Revolutionary Technology of Computer-Generated Artificial Worlds - and How It Promises and Threatens to Transform Business and Society*, New York: Summit, 1991

Rinpoche, Sogyal. *The Tibetan Book of the Living and Dying*, San Francisco: Harper San Francisco, 1994

Robbin, Anthony. "Smithson's Non-Site Sights," *Art News*, Feb. 1969, pp. 50-53

Rose, Barbara. "Art as Experiment, Environment, and Process," *Pavilion*, ed. Billy Klüver/Julie Martin/Barbara Rose, New York: E.P. Dutton, 1970, pp. 60-104

Rose, J. *The Cybernetic Revolution*, London: Elek Science, 1974

*Rosen, Nancy. *Ten Years of Public Art, 1972-1982*, New York: Public Art Fund, 1982

Ross, David A. "Nam June Paik's Videotapes," *Nam June Paik*, New York: Whitney Museum of American Art, 1982, pp. 101-110

*Roszak, Theodore. *The Making of a Counter Culture*, Garden City, NY: Anchor, 1969

Rucker, Rudy. *R.U. Sirius and Queen Mu, Mondo 2000: A User's Guide to the New Edge*, New York: Harper Perennial, 1992

Rushkoff, Douglas. *Cyberia: Life in the Trenches of Hyperspace*, San Francisco: Harper, 1994

Ryan, Paul. "Cable Television: The Raw and the Overcooked," *Radical Software*, No. 1, Summer 1970, p. 12

Ryan, Paul. "Cybernetic Guerilla Warfare," *Radical Software*, No. 3, Spring 1971, pp. 1-2

Ryan, Paul. "Attempting a Calculus of Intention," *Radical Software*, Vol. 1, No. 3, 1971

Ryan, Paul. "Toward an Information Economy," *Radical Software*, Vol. 1, No. 3, 1971

Ryan, Paul. *Cybernetics of the Sacred*, Garden City, NY: Anchor Books, 1974

Ryan, Paul. "The Ecochannel Design," *IS Journal*, Vol. 2, No. 2, 1987, pp. 46-64

Ryan, Paul. "A Genealogy of Video," *Leonardo*, Vol. 21, No. 1, 1988, pp. 39-44

Ryan, Paul. "The Earthscore Notational System for Orchestrating Perceptual Consensus about the Natural World," *Leonardo*, Vol. 24, No. 4, 1991, pp. 457-465

Ryan, Paul. *Video Mind, Earth Mind. Collected Writings*, New York: Peter Lang Publishing, 1992

Ryan, Paul. "The Millenium, Montage, Remote Sensing and the Earthscore Method," *Codes Customs. Milennial Perspectives*, ed R. Kevilson, New York: Peter Lang Publishing, 1994, pp. 245-267

Saad-Cook, Janet. "Touching the Sky: Artworks using Natural Phenomena, Earth, Sky, and Connections to Astronomy,", *Leonardo*, Vol. 21, No. 3, 1988, pp. 123-134

*Schneider, Ira and Beryl Korot. eds., *Video Art: An Anthology*, New York: Harcourt Brace Jovanovich, 1976.

Scott, Gail. "Newton Harrison," A *Report on the Art and Technology Program*, Los Angeles: Los Angeles County Museum of Art, 1971, pp. 118-126

*Sennett, Richard. Flesh and Stone. *The Body and the City in Western Civilization*, London/Boston: Faber and Faber, 1994

Shamberg, Michael. "Guerilla Television," *Radical Software*, Vol. 1, No. 6, 1971

Shannon, Claude E., "A Mathematical Theory of Information," *Bell System Technical Journal*, No. 27, 1948, pp. 632-656

Sharp, Willoughby. "Luminism and Kineticism," *Minimal Art*, ed. G. Battcock, New York: E.P. Dutton, 1968, pp. 317-358

Sharp, Willoughby. "From Teenage Monk to Video Bareback Rider. Interview with Paul Ryan," *Video 81*, Vol. 2. No. 1, 1981, pp. 14-17

*Sheldrake, Rupert, *A New Science of Life: The Hypothesis of Formative Causation*, Los Angeles: J.P. Tarcher, Inc., 1981

Sherk, Bonnie. "The Farm," *Social Works*, Los Angeles: Los Angeles Institute of Contemporary Art (LAICA), 1979, pp. 38-40

Shirey, David L. "Video Art Turns to Abstract Imagery," *The New York Times*, July 4, 1972

Siegel, Jeanne. "An Interview With Hans Haacke," *Arts Magazine*, May 1971, pp. 18-21

Silver, K.E. "Nam June Paik: Video's Body," *Art in America*, Nov. 1993, pp. 100-107

Sky, Alison. "Entropy Made Visible. Interview With Robert Smithson," *On Site*, No. 4, 1973, pp. 188-199

Smith, Roberta. "Robert Irwin: The Subject is Sight," *Art in America*, March/April 1976, pp. 68-73

Smith, Roberta. "Matter Turned Into Light and Space," *The New York Times*, 25 September 1992, Section C, p. 29

Smithson, Robert. "Entropy and the New Monuments," *Artforum*, June 1966, pp. 26-31

Smithson, Robert. "Quasi-Infinities and the Waning of Space," *Arts Magazine*, Nov. 1966, pp. 28-31

Smithson, Robert. "Some Void Thoughts on Museums," *Arts Magazine*, Feb. 1967, p. 41

Smithson, Robert. "Towards the Development of an Air Terminal Site," *Artforum*, June 1967, pp. 35-40

Smithson, Robert. "A Sedimentation of the Mind: Earth Projects," *Artforum*, Sept. 1968, pp. 44-50

Smithson, Robert. "Frederick Law Olmsted and the Dialectical Landscape," *Artforum*, Feb. 1973, pp. 62-68

Smithson, Robert. *The Writings of Robert Smithson*, ed.Nancy Holt, New York: New York University Press, 1979

Smithson, Robert. "The Spiral Jetty," *Arts of the Environment*, ed. G. Kepes, New York: George Braziller, 1972, pp. 109-116

Smithson, Robert. "Interpolation of the Enantiomorphic Chambers," *The Writings of Robert Smithson*, 1979, p. 39

Snow, Charles Percy. *The Two Cultures*, New York: New American Library, 1964

Sonfist, Alan. *Natural Phenomena as Public Monuments*, Purchase, NY: Neuberger Museum/State University of New York, 1978

Sonfist, Alan. ed., *Art in the Land*, New York: E.P. Dutton, 1983

Sontag, Susan. *Against Interpretation*, New York: Farrar, Strauss & Giroux, 1967

Spruch, G.M. "Report on the Symposium on Art and Science Held at the Massachusetts Institute of Technology," *Artforum*, Jan. 1969, p. 28

Steina. "My Love Affair with Art: Video and Installation Work," *Leonardo*, Vol. 28, No. 1, 1995, pp. 15-18

Steina. "Reflections," *Steina e Woody Vasulka. Video, media e nuove immagini nell' arte contemporanea.* Rome: Palazzo delle Esposizioni, 1995, pp. 98-105

Stiles, Kristine. "Helen and Newton Harrison: Questions," *Arts Magazine*, Feb. 1978, pp. 131-133

Stonier, Tom. *Information and the Internal Structure of the Universe*, London: Springer, 1990

Sturken, Marita. "The Videotapes of Bill Viola," *Afterimage*, Summer 1982, pp. 28-30

Sturken, Marita. "Steina and Woody Vasulka: In Dialogue with The Machine," *Steina and Woody Vasulka. Machine Media*, San Francisco: San Francisco Museum of Modern Art, 1996 pp. 35-48

Toffler, Alvin. *Future Shock*, New York: Random House, 1970

Tomkins, Calvin. "Onward and Upward With the Arts: E.A.T.," *The New Yorker*, October 3, 1970

Tomkins, Calvin. "Onward and Upward With the Arts. Maybe a Quantum Leap," *The New Yorker*, February 5, 1972, pp. 42-67

Tomkins, Calvin. "Outside Art," *Pavilion*, ed. Billy Klüver/Julie Martin/Barbara Rose, New York: E.P. Dutton, 1972, pp. 105-165

Tomkins, Calvin. "Nam June Paik: Video Visionary," *The New Yorker*, May 5, 1975, pp. 44-79

Tomkins, Calvin. *Off the Wall*, New York: Penguin Books, 1980

Topper, David R. and John H. Holloway. "Interrelationships between the Visual Arts, Science and Technology," *Leonardo*, Vol. 13, No. 1, 1980, pp. 29-33

Topper, David R. and John H. Holloway. "Interrelationships between the Visual Arts, Science and Technology, A Bibliographic Up-date, " *Leonardo*, Vol. 18, No. 3, 1985, pp. 197-200

Tuchman, Maurice. ed., *A Report on the Art and Technology Program*, Los Angeles: Los Angeles County Museum of Art, 1971

Vasulka, Woody. "A Syntax of Binary Codes," *Afterimage*, Summer 1978, p. 20

Vasulka, Woody. "Brotherhood - Table III," *ars electronica*: Linz, Austria, 1994, pp. 87-90

Vasulka, Woody. "Notes on Installations: From Printed Matter to Noncentric Space," *Steina and Woody Vasulka. Machine Media*, San Francisco: San Francisco Museum of Modern Art, 1996, pp. 65-72

Vasulka, Woody. "Toward a non-centric narrative space," *Video, media e nuove immagini nell' arte contemporanea*. Rome: Palazzo delle Esposizioni, 1995, pp. 106-109

Vinklers, Bitite. "Art and Information," *Arts Magazine*, Sept./Oct. 1970, pp. 47-49

Viola, Bill. "Video as Art," *Video Systems Magazine*, Vol. 8, No. 7, July 1982, pp. 26-35

Viola, Bill. "Will There be Condominiums in Data Space?," *Video 80*, No. 5, Fall 1982, pp. 36-41

Viola, Bill. "History, 10 Years, and the Dreamtime," *Video: A Retrospective 1974-1984*, ed. Kathy Huffman, Long Beach, CA: Long Beach Museum of Art, 1984, pp. 18-23

Viola, Bill. "Excerpt of the Sound of One Line Scanning," *1986 National Video Festival*, Los Angeles: American Film Institute, 1986, pp. 40-44. Printed in full in: *Sound by Artists*, eds. Dan Lander and Micah Lexier, Toronto: Art Metropole and Banff: Walter Philips Gallery, 1990, pp. 39-54

Viola, Bill. "Video Black - The Mortality of the Image, " *Illuminating Video: An Essential Guide to Video Art*, eds. Doug Hall and Sally Jo Fifer, San Francisco: Aperture/Bay Area Video Coalition, 1990, pp. 476-468

Viola, Bill. "Perception, Technology, Imagination, and the Landscape," *Enclitic*, Vol. 11, No. 3, July 1992, pp. 57-60

Viola, Bill. "On Transcending the Water Glass," *CyberArts: Exploring Art and Technology*, ed. Linda Jacobson, San Francisco: Miller Freeman Inc., 1992, pp. 3-5

Viola, Bill. *Reasons for Knocking at an Empty House: Writings 1973-1994*, ed. Robert Violette, Cambridge, MA: MIT Press and London: Thames and Hudson, 1995

*Waddington, C.H. *Behind Appearance*, Cambridge, MA: MIT Press, 1970

Wasserman, Emily. "Explorations - Toward a Civic Art," *Artforum*, June 1970, pp. 87-89

Wechsler, Judith. "György Kepes," *György Kepes: MIT Years 1945-1977*, Cambridge, MA: MIT Press, p. 10

Weschler, Lawrence. "Robert Irwin's Whitney Project - Retrospects and Prospects," *LAICA Journal*, No. 15, July/Aug. 1977, pp. 14-23

Weschler, Lawrence. *Seeing is Forgetting the Name of the Thing One Sees: A Life of Contemporary Artist Robert Irwin*, Berkeley: University of California Press, 1982

Weschler, Lawrence. "Playing It as It Lays & Keeping It in Play: A Vist with Robert Irwin," *Robert Irwin*, New York/Los Angeles: Rizzoli International Publications and The Museum of Contemporary Art, 1993, pp. 153-190

Whitman, Simone. "Theater and Engineering - An Experiment," *Artforum*, Feb. 1967, pp. 26-33

Wick-Kmoch, Astrid. "Kunst + Systemtheorie + Sozialwissenschaften: Zu den Arbeiten von Hans Haacke," *Kunstforum International*, Band 27, 1978, pp. 125-142

Wiener, Norbert. *Cybernetics or: Control and Communication in the Animal and the Machine*, New York: John Wiley & Sons, 1948

Wiener, Norbert. *The Human Use of Human Beings*, Boston, MA: Houghton Mifflin, 1950
*Williams, Raymond. *Television: Technology and Cultural Form*, New York: Schocken Books, 1974
Yalkut, Jud. "Art and Technology of Nam June Paik," *Arts Magazine*, April 1968, pp. 50-51
Yalkut, Jud. "Sao Paulo, No! Smithsonian, Yes!," *Arts Magazine*, May 1970, pp. 16-18
Yalkut, Jud. "Conversations With György Kepes," *Arts Magazine*, May 1970, pp. 16-18
Youngblood, Gene. "World Game: the artist as ecologist," *Artscanada*, Aug. 1970, pp. 42-49
Youngblood, Gene. "The Open Empire," *Studio International*, April 1970, pp. 177-178
Youngblood, Gene. *Expanded Cinema*, New York: E.P. Dutton, 1970
Youngblood, Gene. "Buckminster Fuller's World Game," *Whole Earth Catalog*, March 1970
Youngblood, Gene. "Die neue Renaissance: Kunst, Wissenschaft und die universelle Maschine," *ars electronica 1984*, Linz ,1984.
Youngblood, Gene. "Metadesign - Die neue Allianz und die Avant-garde," *Kunstforum International*, Band 98, Jan./Feb. 1989
Zeitlin, Marilyn A. "Continuities in the Work of Bill Viola," *Bill Viola: Buried Secrets*, Tempe, AZ: Arizona State University Art Museum, in association with the Kestner-Gesellschaft, Hannover , Germany, 1995, pp. 53-63

Index